Statistical Evaluation of Measurement Errors

Design and Analysis of Reliability Studies

Second Edition

Graham Dunn
University of Manchester

John Wiley & Sons, Ltd

First published in Great Britain in 1989
Second edition published in 2004 by
Hodder Arnold, a member of the Hodder Headline Group,
338 Euston Road, London NW1 3BH

Reprinted by John Wiley & Sons 2009

John Wiley & Sons Ltd, The Atrium, Southern Gate, Chichester, West Sussex, PO19 8SQ, United Kingdom

For details of our global editorial offices, for customer services and for information about how to apply for permission to reuse the copyright material in this book please see our website at www.wiley.com.

British Library Cataloguing in Publication Data
A catalogue record for this book is available from the British Library

Library of Congress Cataloging-in-Publication Data
A catalog record for this book is available from the Library of Congress

ISBN 978 0 470 68215 9

1 2 3 4 5 6 7 8 9 10

Typeset in 10/12 pts Times by Charon Tec Pvt. Ltd, Chennai, India

To Jane, Beryl and Tom

Contents

Preface

The statistical methods used to evaluate and compare different methods of measurement are a vital common component of all methods of scientific research, from sociology and psychology, through the biological and medical sciences, to chemistry, physics and precision engineering. Although statistical techniques to investigate the characteristics of measurement errors have often been developed by scientists working within these disciplines, with their specific problems in mind, many of the resulting statistical models have applications far beyond those for which they were originally envisaged. Often, in fact, they have been developed independently within various fields of application without much in the way of exchange of ideas from one discipline to another. Here I try to bring these disparate strands together and to show how much the different approaches have in common. I provide a practically orientated guide to the statistical evaluation of measurement errors, giving particular emphasis to a wide variety of illustrative examples from across the sciences. After first setting the scene by introducing concepts such as precision, reproducibility and reliability, for example, I provide a detailed discussion of the sources of variability of measurements and associated variance components models. The central chapters (i.e. Chapters 3 and 4) deal with the design and analysis of method comparison studies (concentrating on quantitative measurements), ranging from simple paired comparisons to more complex studies involving three or more methods. Much of the statistical methodology here is the extension of the classical linear structural model for pairs of measures, to give models describing linear structural relations amongst several methods of measurement (covariance structure modelling). These chapters lead on to a review of methods for categorical (primarily binary) measures, but again concentrating on statistical models for method-comparison (latent class analysis).

Acknowledgements

Several colleagues have read various draft chapters and made useful suggestions for their improvement. These include Andrew Pickles, Sophia Rabe-Hesketh, Chris Roberts and Anders Skrondal. I am particularly grateful to Andrew and Sophia for initiating me into the use of *gllamm*. I would also like to pay particular thanks to an email correspondent from distant Hawaii, Thomas Knapp. Tom first contacted me when I was struggling with the first draft of Chapter 1. By then I had already passed the first of many deadlines for submission of the whole manuscript to the publishers. It was Tom's helpful comments on Chapter 1, together with his encouragement and enthusiasm, which motivated me to get on with the rest of the job. Even then, it has still taken me a lot longer to complete the text than I had originally envisaged, so I would like to thank the commissioning editor at Arnold, Liz Gooster, for her patience of the last year or two. Finally, I would like to thank my wife Jane for her patience, particularly over the last few months of the project, as I spent much more of my supposedly free time at the keyboard than I should have done!

A major component of the present book is the illustration of the statistical ideas through extensive analysis of data sets from a variety of scientific disciplines. I would like to thank my colleagues, Professor Glyn Lewis and Dr Stuart Turner, for letting me use data from the days when we all three worked at the Institute of Psychiatry in London (measurements of psychological distress and ventricle-brain ratios, respectively). Other data sets are from the published literature. Both I and the publishers would like to thank the following for using copyright material in this book:

Blackwell Publishers (Table 3.6)
The American Statistical Association (Table 4.23)
The Australasian Medical Publishing Company (Table 4.5)
The Biometrical Journal (Table 4.9)
The Canadian Anesthesiologists' Society (Table 2.2)
The International Biometric Society (Table 4.1, 4.14 and 5.22).

1

Setting the scene

1.1 Introduction

Imagine a rather obsessional cook who decides to test the 'accuracy' of his kitchen scales. The basic features of these scales incorporate a large pan on the top and a pointer and scale beneath to indicate the weight of the contents of the pan. The only other feature of interest is a lever which enables him to re-set the pointer's zero position when the pan contains nothing. The scale indicates weights from 0 to 3 kg in 20 g intervals. It is unlikely, therefore, that the cook could determine the value of a single reading any closer than to the nearest 10 g. He takes a packet of dried fruit from his kitchen cupboard and notes that the packers have claimed that it weighs 500 g. This will be assumed, for the time being, to be the packet's true weight. He takes ten measurements of the fruit's weight on his scales and obtains the following readings (in g):

490, 520, 500, 520, 470, 490, 490, 490, 500, 480

The arithmetic mean of these recordings is 495 and their variance is 250 (equivalent to a standard deviation of 15.8). The arithmetic mean gives an indication of the weight that would be obtained, on average, if the weighing process could be repeated indefinitely, and the difference between this value and the packet's 'true' weight of 500 g indicates, on the basis of the results of this simple experiment at least, that the scales seem to underestimate the weight of an item of food. The scales have an estimated bias of −5 g. The variance (or, equivalently, the standard deviation) of the readings gives the cook an idea of the precision of the scales. A high variance is an indicator of low precision.

Now, having thought about the results of the first experiment, the cook realises that the bias observed may have been caused by the way the zero had been set prior to the experiment. It could easily have been due to chance, of course, but, in any case, he decides to carry out a further experiment. This time he weighs the same packet of dried fruit ten times, but prior to each weighing he re-sets the machine's pointer to zero. The ten readings now obtained are the following:

480, 500, 510, 450, 480, 500, 490, 460, 480, 450

Their arithmetic mean is 480 (i.e. the bias is now estimated to be $-20\,g$) with variance 444.4 (standard deviation 21.1).

Although we cannot have too much confidence in the actual results, these two experiments are useful in illustrating the concepts of bias and precision. They also illustrate that both the bias and precision are dependent on the way the measuring instrument is being used, and not solely on the identity of the particular weighing machine. A particular source of error (such as re-setting the zero on the kitchen scales) can contribute to bias or to lack of precision. In the cook's first experiment the zero was set once only and therefore the influence of this setting was common to all readings (that is, it is a potential source of bias). In the second experiment the same re-setting operation was carried out before each reading was taken, and in this case the procedure resulted in potentially lower precision. A systematic error in the first experiment has become a random or haphazard one in the second. The apparent change in bias, however, cannot be explained by the change in experimental design, and larger and possibly more complex experiments would be needed in order to evaluate all of the various possible sources of error.

Having learnt a little about errors of measurement, how should the cook proceed to use his kitchen scales? Higher precision can be gained by just setting the scales' zero once. This may or may not be a good procedure depending on how easily or how quickly the zero drifts away from its pre-set position during use. This does not solve the problem of bias, however. This might be solved by first weighing an empty container and then adding the required amount of food. In this case the bias ought to be common to both readings and would disappear on subtraction to obtain the weight of the food alone. The improvement might be an illusion, however. The variance of the difference will be double that of the variance of an individual reading (see Section 1.2, below). However, the new problem can be solved quite easily by repeating the whole measurement process several times and then calculating the mean of the differences obtained on each replication.

Before leaving the kitchen for the scientific laboratory it might be useful to consider the implications of the errors of measurement introduced by the use of the above kitchen scales. The conclusion ought to be that the likely size of the errors introduced by these scales when weighing foods such as sugar, apples or potatoes is trivial. A 25 g error should be of no practical consequence to the cook, however obsessional he might be. This can be illustrated by introducing the coefficient of variation (c.v.) as a measure of relative precision. The sample c.v. is simply obtained by dividing the standard deviation of the measurements by the corresponding arithmetic mean. The coefficients of variation obtained from the two experiments above are 0.03 and 0.04, respectively (if preferred, the c.v. can be multiplied by 100 and be expressed as a percentage). But what if the true weight were 5 g instead of 500 g? And what if the standard deviations of the measurements were to remain as before? The new c.v. would be about 0.3 or 30%. If the cook were to weigh very small amounts of spices, for example, then the above standard deviations would no longer be trivial. The difference between 1 g and 30 g of hot chilli powder when added to a stew is enormous! Clearly one needs scales of much higher precision for the measurement of commodities such as spices. In other words, the accuracy required of any measurement instrument is dependent on the context of its use; a certain lack of precision (or bias) may be trivial in one context but absolutely vital in another. If you are not convinced, consider the differences in approach needed for monitoring and auditing the sale and transport of scrap iron and those of weapons-grade uranium. No one minds if a few pounds of scrap iron get lost, but we are desperately concerned that we can account for all of the uranium.

1.2 Statistical models for physical measurements

When we move from the kitchen to a physics or chemistry laboratory, we wish to go further than the rough and ready analysis presented above. Ideally, we should be able to understand the measurement process well enough to be able to make inferences about the likely errors in any given measurement or estimate of a given characteristic (bias or precision, for example). What is its standard error, for instance? How confident can we be that the precision falls within a given range? How many repeated measurements need to be made to achieve an estimate with a given standard error? How many measurements are needed to convince a sceptic that the bias has an absolute value that is less than some pre-determined requirement?

It is a truism in physics that a measurement or estimate is worthless unless it is accompanied by an estimate of its precision (in the form of a range of plausible values, a confidence interval or simply its standard error, for example). To illustrate this point, Taylor (1982) describes a problem similar to one that is said to have been solved by Archimides. We have to decide whether a crown is made of 18-carat gold (with a known density of $15.5 \, g/cm^3$) or a cheap alloy (with a known density of $13.8 \, g/cm^3$). Two independent estimates of the crown's density are $13.9 \, g/cm^3$ and $15 \, g/cm^3$. On their own we cannot tell which of the two estimates is the more reliable, nor can we make the required decision using these data. The data as they stand are worthless. If, however, more information were to be provided, then we might be able to make use of these estimates. The expert who provided the first of the estimates also stated that it is highly probable (95%, say) that the interval $13.7 \, g/cm^3$ to $14.1 \, g/cm^3$ includes the true value. The expert providing the second estimate stated that he thought that the true density must lie between $13.5 \, g/cm^3$ and $16.5 \, g/cm^3$. Providing that these two experts could justify their conclusions concerning the above intervals (and, in practice, they would need to) then we can justifiably conclude that the crown is made of alloy.

The simplest statistical model for a series of measurements of weights or densities has the following form:

$$X_{ij} = \tau_i + e_{ij} \tag{1.1}$$

In this model, X_{ij} represents the jth measurement on the ith object ($i = 1, 2, ..., I$; $j = 1, 2, ..., J$), τ_i is the 'true' value for object i, and e_{ij} is the measurement error associated with X_{ij}. We assume that the mean of the measurement errors over repeated measurements is equal to zero. That is,

$$E_j(e_{ij}) = 0 \tag{1.2}$$

or, equivalently,

$$E_j(X_{ij}) = \tau_i \tag{1.3}$$

Note also that

$$E_{i,j}(X_{ij}) = E_i(\tau_i) = \mu \tag{1.4}$$

where μ is the mean of the population of possible measurements, and in all three of these equations the subscripts for the expectation operators refer to the population of values over which the expectations are being taken. Note that we are defining the true value in terms of an expectation, rather than in reference to an external standard as in the case of the earlier $500 \, g$ of dried fruit. An alternative interpretation is that we have assumed bias to be zero. If the measurement procedure is, indeed, biased then the bias is confounded with the τ_is. We will return to the joint problems of truth, bias and standards later in this chapter (see Section 1.5). Staying with our simple model, the error terms (e_{ij}) are usually assumed to be statistically independent of one another (both between and within objects) and to be independent of the true values (τ_i). These assumptions imply that

$$\sigma_X^2 = \sigma_\tau^2 + \sigma_e^2 \tag{1.5}$$

where σ_X^2 is the variance of the observations, σ_τ^2 is the variance of the true values, and σ_e^2 the variance of the measurement errors. Assuming that the measurement conditions are being held constant, the precision of the measurement process is measured by the square root of σ_e^2, which is called the repeatability standard deviation (International Standards Institute, 1994a), or more simply, repeatability. An equivalent term, more often used in the social and behavioural sciences, is the standard error of measurement (sem – see Section 1.6 for more details).

Now let's complicate matters by considering a measurement study in which all of the above IJ measurements are repeated on two separate occasions. Let's also assume that the measurement process has changed slightly from the first occasion to the next (the operator may have changed, or the instrument re-calibrated, for example). For occasion 1 we have:

$$X_{ij1} = \tau_{i1} + e_{ij1} \tag{1.6}$$

and, for occasion 2:

$$X_{ij2} = \tau_{i2} + e_{ij2} \tag{1.7}$$

The use of the subscripts 1 and 2 should be clear from the context. Note that, because we have defined true values as expectations, it is possible for τ_{i1} and τ_{i2} to differ (and, of course, for $\mu_1 \neq \mu_2$). The data from such a study enable us to estimate $2I$ means and $2I$ standard deviations, from which we can infer patterns of variation in bias and precision both within and between occasions. An alternative design might involve each of the IJ measurements being repeated in two or more, say, K laboratories. However, instead of choosing objects at random, in the physical sciences it is more usual to select a range of standards (fixed 'known' amounts of material or concentrations of chemicals, for example) to span the operational range of the measurement procedure in routine practice. Such an experiment is called an inter-laboratory precision study (see Sections 1.6, 4.9 and 4.11). The former design (selecting subjects or objects to be measured at random from a defined target population) is more typical of the social scientists' generalizability study (see Section 1.6). Here we return to the cook in his kitchen and to a rather simplified version of the two occasion study in which the objects to be measured might be considered as a random sample from a larger population. Note that in this situation the cook has no *a priori* knowledge of the weights of the objects. There is no 'true' weight marked on the side of the packet.

Table 1.1 shows measurements obtained in a third, slightly more complicated, study with the kitchen scales. Here the scales' pointer was set to zero at the start of the study. Then 20 objects (packets of potatoes) were each weighed separately. The pointer on the scales was then again set to zero (in case of any instrumental drift) and the 20 packets each weighed a second time. The statistical model for this experiment is provided in Equations (1.6) and (1.7). Here, since there is only one observation for each time point (i.e. $j = 1$ for every one) the j is dropped from the subscripts. But we have a slight problem. We cannot estimate each of the differences $(\tau_{i1} - \tau_{i2})$ without assuming that all of the measurement errors are zero. They are confounded. However, if it is assumed that

$$\tau_{i1} - \tau_{i2} = \alpha_{12} \tag{1.8}$$

Table 1.1 Repeated measurements of the weights of several packets of potatoes (in grammes).

Packet	Time 1	Time 2	Difference
1	560	550	10
2	140	140	0
3	1680	1710	−30
4	1110	1090	20
5	1060	1040	20
6	280	250	30
7	620	600	20
8	830	800	30
9	690	690	0
10	1210	1170	40
11	2300	2260	40
12	1880	1850	30
13	2000	2000	0
14	2360	2300	60
15	1670	1720	−50
16	1230	1230	0
17	1370	1390	−20
18	1750	1710	40
19	1730	1710	20
20	1680	1640	40
		Mean	15
		Variance	721

for all i, then equations (1.6) and (1.7) can be replaced by

$$X_{i1} - X_{i2} = \alpha_{12} + (e_{i1} - e_{i2}) \tag{1.9}$$

The constant relative bias, α_{12}, can be estimated from the mean of the $X_{i1} - X_{i2}$. If we assume that the error variances are the same on the two occasions then it can be estimated from half the variance of the $X_{i1} - X_{i2}$. This follows from

$$\begin{aligned}
\text{Var}(X_{i1} - X_{i2}) &= \text{Var}(e_{i1} - e_{i2}) \\
&= \text{Var}(e_{i1}) + \text{Var}(e_{i2}) \\
&= 2\sigma_e^2
\end{aligned} \tag{1.10}$$

As before, we have assumed that the errors are all statistically independent, both of each other and of the other model parameters.

1.3 Latent variable models

Table 1.2 shows four measurements on each of 15 pieces of string. The first column gives their lengths as measured using a scaled rule marked with divisions 1/10th inch apart. The other three columns give the lengths of these pieces of string independently estimated (guessed!) by three observers without the use of the rule. How do we approach the analysis of these data? One possibility is to regard those lengths provided by the scaled rule as a standard (S), regarded to be the truth. We can then plot and fit a simple calibration curve for each of the sets of guesses (Graham, Brian & Andrew) against the standard (S). A plot of Graham's guesses against S is shown, for example, in Figure 1.1. We can fit a simple linear

Table 1.2 Subjective estimates of the lengths of 15 pieces of string (to nearest 1/10 inch).

String	Rule	Graham	Brian	Andrew
A	6.3	5.0	4.8	6.0
B	4.1	3.2	3.1	3.5
C	5.1	3.6	3.8	4.5
D	5.0	4.5	4.1	4.3
E	5.7	4.0	5.2	5.0
F	3.3	2.5	2.8	2.6
G	1.3	1.7	1.4	1.6
H	5.8	4.8	4.2	5.5
I	2.8	2.4	2.0	2.1
J	6.7	5.2	5.3	6.0
K	1.5	1.2	1.1	1.2
L	2.1	1.8	1.6	1.8
M	4.6	3.4	4.1	3.9
N	7.6	6.0	6.3	6.5
O	2.5	2.2	1.6	2.0

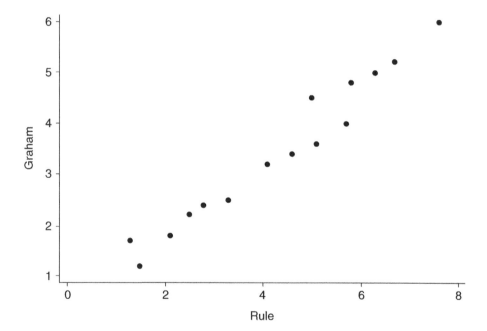

Figure 1.1 String lengths: Graham versus the Rule.

regression using least squares and estimate the slope of the line together with the standard deviation of the residuals (measurement errors), assuming, perhaps naively, that the variability of the residuals is constant and not dependent on the value of S. We will not report the detailed analyses here (see Sections 2.5.3, 4.6 and 4.7), but it is clear from Figure 1.1 for example, that the lengths of the pieces of string are consistently being underestimated by Graham and that this underestimation is greater for the longer pieces of string (i.e. the regression coefficient in our simple regression model is less than 1).

Let X_{ik} be the length of the ith piece of string as estimated by the kth observer ($i = 1$ to 15; $k = 1, 2, 3$). Let S_i be the length of the ith piece of string as determined by the scaled rule. A general model that can be used to relate X_{ik} to S_i is

$$X_{ik} = \alpha_k + \beta_k S_i + e_{ik} \tag{1.11}$$

with the usual assumptions concerning the independence of the measurement errors.

Now, what happens if we either do not have access to a standard, S, or we are not prepared to accept that it is error free – that is, we are not prepared to accept that S is the truth? A straightforward solution is to postulate that a true value, τ, exists, but that we are not able to measure it. τ is explicitly a latent or hidden variable. Discarding S from the data, our model now becomes

$$X_{ik} = \alpha_k + \beta_k \tau_i + e_{ik} \tag{1.12}$$

But, instead of discarding S we could add an extra equation as follows:

$$S_i = \alpha_s + \beta_s \tau_i + e_{is} \tag{1.13}$$

where the interpretation of the parameters should be obvious from the context. Fitting such a model to either the data from the three observers, or to the observers plus the rule, will be left to Sections 4.6 and 4.7. As well as expecting perhaps all of the α_k but particularly α_s to be close to zero, we would expect β_s to be very close to 1 (if not actually equal to 1) and σ_{es}^2 to be considerably smaller (perhaps by an order of magnitude) than the σ_{ek}^2. Here σ_{es}^2 and σ_{ek}^2 refer to the variance of the errors in S and observer k, respectively.

Although we have approached the problem in a slightly different way to that in earlier sections, there is really nothing particularly new here. Our interpretation of the simple model described by Equations (1.6), (1.7) and (1.8) implied that

$$\tau_{i1} = \alpha_{12} + \tau_{i2} \tag{1.14}$$

Our more general latent variable model implies that

$$\tau_{i1} = \alpha_1 + \beta_1 \tau_i \tag{1.15}$$

and

$$\tau_{i2} = \alpha_2 + \beta_2 \tau_i \tag{1.16}$$

It follows that

$$\tau_{i1} = \alpha_{12} + \beta_{12} \tau_{i2} \tag{1.17}$$

where α_{12} is $\alpha_1 - \alpha_2 \beta_{12}$, and β_{12} is the ratio β_2/β_1. The correlation between the true scores of the two tests is perfect (that is, unity). In psychometric test theory any two tests (that is, measurement procedures) which have this property are known as congeneric tests (Jöreskog, 1971a). If the βs are same for two tests then they are known essentially as τ-equivalent. If the true values are equal then they are τ-equivalent in a more strict sense and, finally, if their true values are equal and they have the same precision they are said to be parallel tests. So, for example, applying this same terminology to physical measurements, two thermometers, one scaled in °C and the other in °F, could be regarded as congeneric. Two thermometers, both well calibrated in °C, could be either τ-equivalent or parallel. In the first case they would have unequal precisions (an expensive well-made device as opposed to a cheap one) and in the second they would have the same precision (they could be instruments of the same make and from the same batch of manufacture, for example).

1.4 Psychometric tests: the factor analysis model

The three subjective judgements of the lengths of pieces of string given in Table 1.2 are highly correlated. These correlations result from the measurement model described by Equation (1.11). In this example, we can be reasonably confident that the measurement model is more-or-less valid because the judgements were all made independently of one another, and also because we are all very familiar with the concept of length. It is also help-ful that we have access to measurements we can consider the true lengths – the standard determined by the scaled rule. In the measurement of psychological or social characteris-tics, however, it is the case that neither do we have a measurement model such as Equation (1.1) that is intuitively obvious, nor do we have access to any standards that we might regard as the truth. There is no equivalent to the scaled rule in the behavioural or social sciences. However, we can infer a measurement model such as (1.11) from the observed pattern of correlation between the fallible indicators of the characteristic we are trying to measure. Here the measurement model is introduced to 'explain' or account for the correlations observed. The measurement model that is most widely known in this area is the linear factor model first proposed by Spearman in 1904 (see Bartholomew, 1995; or Bartholomew & Knott, 1999).

Consider a battery of cognitive tests that can be used to measure intellectual ability. These could, for example, be examinations in school subjects such as mathematics, classics, English and history (Spearman, 1904) or the subtests of a modern intelligence test (Wechsler, 1981). Let Y_{ik} represent the score obtained by the ith subject on the kth test ($i = 1, 2, ..., I$; $k = 1, 2, ..., K$). Spearman's *common-factor model* postulates that each test score is made up of two components in the following way:

$$Y_{ik} = \lambda_k f_i + u_{ik} \tag{1.18}$$

where, because of the arbitrary scaling of psychological measurements, both the Y_{ik} and f_i have been standardised to have zero mean and unit variance. The term f_i is a hypothetical true score for the ith individual on an underlying latent scale of cognitive ability (the com-mon factor). The parameter λ_k is the slope of the straight line relating test k to the common factor. It is equivalent to the βs of our earlier models but in the literature on factor analysis is usually called a factor loading. The last term in the model, u_{ik}, is known as a specific factor, and represents that part of the observed test score that is not accounted for by the common factor. Unlike Equation (1.12), there is no intercept term in this model because we have arbitrarily fixed the mean and variance of the factor scores (that is, the f_i) to have a zero mean and unit variance.

From (1.18) we have

$$\begin{aligned} \text{Var}(Y_{ik}) &= \lambda_k^2 \text{Var}(f_i) + \text{Var}(u_{ik}) + 2\lambda_k \text{Cov}(f_i, u_{ik}) \\ &= \lambda_k^2 + \sigma_{uk}^2 \end{aligned} \tag{1.19}$$

where σ_{uk}^2 is the variance of the u_{ik} and on the assumption that the factor variance is 1 and that f_i and u_{ik} are uncorrelated. If, as indicated above, the test scores have been standardised to have a mean of 0 and variance of 1, then

$$1 = \lambda_k^2 + \sigma_{uk}^2 \tag{1.20}$$

or

$$\lambda_k^2 = 1 - \sigma_{uk}^2 \tag{1.21}$$

So, λ_k^2 is the proportion of the variance of the kth test that is explained by the underlying common factor. The term communality is used by factor analysts to refer to this proportion. $1 - \lambda_k^2$ is known in the factor analysis literature as the test's specificity (but take care not to confuse this use of the word 'specificity' with its use in the description of the performance of diagnostic tests – see Chapter 5 – we will not use the word in the factor analysts' sense in what follows).

For two standardised test scores, Y_{ik} and Y_{im},

$$\begin{aligned}
\text{Cov}(Y_{ik}, Y_{im}) &= \text{Corr}(Y_{ik}, Y_{im}) \\
&= E(Y_{ik}, Y_{im}) \\
&= E[(\lambda_k f_i + u_{ik})(\lambda_m f_i + u_{im})] \\
&= \lambda_k \lambda_m
\end{aligned} \tag{1.22}$$

To summarise, the expected proportion of the test's variance that is explained by the common factor is the square of the test's loading (λ_k^2). That proportion not so explained is therefore $1 - \lambda_k^2$. The expected correlation between scores on two different tests, k and m, is given by the product of the respective loadings (that is, by $\lambda_k \lambda_m$).

In terms of its mathematical properties, the model inferred in (1.17) is identical to the one for the lengths of pieces of string (1.11) – after appropriate standardisation, of course. The only difference is the way they were arrived at. On the one hand, Equation (1.11) was constructed to describe what we intuitively believed to be the truth but, on the other, Equation (1.17) has been proposed to explain an observed pattern of correlations. We 'know' what length is. It is 'real' and, in principle, at least, we can measure it directly with a great deal of precision. Intelligence, on the other hand, is a much more illusive characteristic. Can we ever describe it as having a real existence outside the model? Can we ever measure it directly, never mind with any degree of precision? No. But can length have a reality outside a given physical model of the world? Possibly, but I'm not convinced. Perhaps it is just a matter of degree rather than a vital qualitative difference, particularly if we acknowledge that no characteristic or construct, whether in the physical or social sciences, can ever be measured completely without error. We will not get too bogged down by philosophical debate here, but follow Bartholomew (1996) in treating all of the measured characteristics in our examples *as if* they existed. One of the main aims of this book is to stress the similarities of the measurement models across the social, biological and physical sciences and to illustrate how a common set of statistical ideas and methods may be used in a wide variety of disciplines and settings.

Before leaving the present section, however, it may be useful to return to Equation (1.18) and think a little more carefully about the nature of the specific factors (represented by u_{ik}). They are not, necessarily, measurement errors, although they are bound to contain at least some error. We will stress again that they are that part of the test score that is not explained by the common factor. There may be something stable here. A particular student might do particularly well in mathematics in the sense that his or her score in a maths exam is much better than would be predicted by their performance in other academic subjects. It is quite conceivable that if that student were to be given a second mathematics examination then he or she would, again, do very well. A better model for this situation would be one with three random components: (1) the students' general intelligence, (2) the students' ability to cope with specific academic subjects (maths, history, and so on) and, finally, (3) measurement error. Hence

$$Y_{ik} = \lambda_k f_i + s_{ik} + e_{ik} \tag{1.23}$$

where the s_{ik} and e_{ik} are the specific abilities and measurement errors, respectively. However, the s_{ik} and e_{ik} can only be separated and their variances individually estimated if we have replication. Each measuring instrument, test or form of examination must be used at least twice on each object or subject under investigation. Whether this is practically possible, of course, depends on the nature of the measurements and other practical circumstances. With J replicates of each of the K measures on each of the I subjects we have

$$Y_{ijk} = \lambda_k f_i + s_{ik} + e_{ijk} \tag{1.24}$$

Similarly, the more general version of (1.12) is

$$X_{ijk} = \alpha_k + \beta_k \tau_i + s_{ik} + e_{ijk} \tag{1.25}$$

Equations (1.24) and (1.25) represent a much richer measurement model than (1.12) or (1.18) and one that, in one form or another, will be returned to many times in the following chapters.

1.5 Precision, bias and accuracy

How would you compare the precision of a thermometer calibrated in °C with one calibrated in °F? How might the cook assess the performance of his kitchen scales, which weigh objects in grammes, with those of his old-fashioned neighbour, which weigh objects in ounces? I hope that it is clear to the reader that we should not naively compare the respective repeatability standard deviations. A particular standard deviation in ounces (say 5 oz) is a very different amount to a standard deviation of 5 g. We approach this problem by either re-scaling (scale conversion) the repeatability of the new scales from grammes to ounces, or by converting the older ones from ounces to grammes. We will do the latter. For instance, the repeatability of the cook's scales has been estimated to be about 20 g. The repeatability of his neighbour's scales has been estimated by her to be 1.4 oz. We know that 1 ounce is approximately 28 g and so the converted repeatability of the neighbour's scales is about 39 g. The cook is justified in preferring his new metric scales.

Now let's have a look at a model such as (1.12). How do we compare the precision of instrument k, say, with that of instrument m? Again, unless β_k is identical to β_m, the instruments are not scaled using the same units (this is another way of conceptualising these relative biases). If the variances of the measurement errors of the two instruments are σ_{ek}^2 and σ_{em}^2 then the corresponding repeatabilities are σ_{ek}/β_k and σ_{em}/β_m, respectively. If we wish to test the differences between the precisions of the two instruments then it is the differences between these two ratios that should be of interest. We now complicate matters and consider the measurement model in Equation (1.25). In most circumstances we now acknowledge that we have two sources of measurement error, the s_{ik} and e_{ijk}. If their variances are σ_{sk}^2 and σ_{ek}^2, respectively, then the total error variance is $\sigma_{sk}^2 + \sigma_{ek}^2$, and the corresponding repeatability standard deviation to be used in comparison to that of instrument m, for example, is therefore $(\sigma_{sk}^2 + \sigma_{ek}^2)^{\frac{1}{2}}/\beta_k$. If you prefer a number to represent precision that increases as the precision goes up, then $\beta_k/(\sigma_{sk}^2 + \sigma_{ek}^2)^{\frac{1}{2}}$ fits the bill, or even $\beta_k^2/(\sigma_{sk}^2 + \sigma_{ek}^2)$. Shyr and Gleser

(1986) define precision as the ratio of the square of the regression coefficient to the error variance – that is, β_k^2/σ_{ek}^2 for model (1.12). Theobald and Mallinson (1978), on the other hand, prefer a measure of relative precision: $\beta_k^2\sigma_\tau^2/\sigma_{ek}^2$, where σ_τ^2 is the variance of the true values, and 'relative' means relative to the variation between the subjects or objects being measured. If we wish to compare the relative precisions of two instruments, say k and m, then we use the ratio

$$\psi = \frac{\beta_k^2\sigma_{em}^2}{\beta_m^2\sigma_{ek}^2} \tag{1.26}$$

Another, related, but much more familiar index of relative precision is the coefficient of reliability,

$$\kappa_k = \frac{\beta_k^2\sigma_\tau^2}{\beta_k^2\sigma_\tau^2 + \sigma_{ek}^2} \tag{1.27}$$

Reliabilities will be discussed in much more detail in later chapters (but see Section 1.7, below).

When we come to the estimation and comparison of systematic biases (the αs in (1.12), for example) we also have to take note of the value of the βs. Unless we are using a ratio scale of measurement (that is, it has a non-arbitrary, fixed, zero) then it is pointless comparing the intercept terms when the slopes differ.

Before moving on, let's return to the very first sentence of this chapter. We mentioned that the cook wished to assess the accuracy of his scales. We put the word 'accuracy' in quotation marks and then more-or-less ignored the word throughout the rest of the discussion. What does it mean? In a general sense we think of the accuracy of an individual measurement as a measure of how close that measurement is to the truth. In terms of our original model for the kitchen scales it is the sum of the bias of the instrument and the individual random measurement error. We cannot, of course, measure it unless we have access to a standard. And in this situation it is simply the difference between the measurement and that standard. Without a standard it becomes a rather ill-defined concept. Without the standard, the instrument's scale of measurement (that is, the joint values of α and β) is arbitrary, so that we can get rid of systematic biases by scale conversion. The accuracy of an individual measurement then becomes the size of that measurement's random error. But we cannot determine what it is, of course.

But the cook asked about the accuracy of his scales, not that of an individual measurement. Accuracy in this sense can be thought of as a characteristic of a given measuring instrument or of a given measurement procedure (the latter use being preferable). If we are pushed, we can define accuracy as the *mean squared error* (MSE) – with the error here being the difference between the standard and the observed measurement (that is, the sum of the bias and the random error). The MSE, therefore, is the average of the squared deviations of the measurements from their appropriate standards. Alternatively, accuracy might be defined as the square root of the MSE, or *root mean squared error*, RMSE. Once we lose the 'safety' of a true value or standard, however, the concept of accuracy becomes redundant. It is then equivalent to precision. We will rarely discuss accuracy below, and, when we do, it will be in a rather relaxed informal way. We will primarily be concerned with measures of precision (the behaviour of the random components of error) or relative bias (mainly the β coefficients; but the αs having some interest).

1.6 Reproducibility and generalizability

The International Standards Organization (ISO) publishes a series of booklets on the conduct of inter-laboratory precision studies – International Standard ISO 5725 (ISO, 1994a to 1994e, 1998) in which it defines the experimental conditions recommended for the determination of precision and bias. The International Standard lists the following factors, amongst others, which contribute to the variability of the results from applying a particular measurement procedure:

(a) the operator
(b) the equipment used
(c) the calibration of the equipment
(d) the environment (temperature, humidity, air pollution and so on)
(e) the time elapsed between measurements.

Our kitchen scales have illustrated several of these sources of variation. The operator could be the cook or his neighbour; the equipment could be the cook's scales or the neighbour's scales; calibration could include the scale of measurement (metric or imperial) together with any adjustment procedures (such as resetting the zero); the environment depends on the conditions within the two kitchens and, finally, a long period between repeated weighings might lead to a slow (or not so slow) drift in the behaviour of the scales. It is not difficult to think of other sources of variability, including the day-to-day changes in the mood of the cook: on some days he's even more obsessional than usual.

ISO 5725 uses the repeatability standard deviation (or, equivalently the repeatability variance or repeatability coefficient of variation) as an index of the precision of a measurement procedure under what it refers to as repeatability conditions. These are conditions under which independent test results are obtained with the same method on identical test items, in the same laboratory, by the same operator, using the same equipment within short intervals of times. In other words, repeatability is the index of precision when the replications of the measurements are made under as near as possible identical conditions.

In contrast to repeatability conditions, the International Standard defines reproducibility conditions as those conditions where test results are obtained with the same method on identical test items, in different laboratories, with different operators using different equipment. The index of precision defined under these conditions is the reproducibility standard deviation (or, equivalently, the reproducibility variance or reproducibility coefficient of variation), or reproducibility for short. Reproducibility gives us a measure of the variability of test results randomly obtained from a variety of possible laboratories. Hopefully, the index of reproducibility will give us a fairly good indication of how laboratory test results might vary in routine practice, but this is not assured. Operators might behave particularly carefully, for example, when they are taking part in a formal precision study.

A similar type of precision study might be carried out on a random sample of specimens to be assayed, rather than on fixed standards as above. I will illustrate using a historically-interesting example of a particularly complex study from clinical chemistry. A paper by MacFarlane *et al.* (1948) describes an extraordinary experiment concerning both method comparisons and sources of observer variation in the determination of haemoglobin. Each of 16 observers made three separate estimations by each of 16 laboratory methods in eight separate samples of blood – a total of 6144 test results! Differences between observers were considered likely to be associated with (a) sex, (b) training, (c) side of the dominant eye, (d) abnormalities of colour vision, and (e) the effect of fatigue. Of the 16 observers taking part in the experiment, eight were men and eight were women. Four men and four women were trained laboratory workers. The remaining observers were relatively unfamiliar with the estimation of haemoglobin. In each of the four above subcategories, two observers were right-eyed and two left-eyed. Finally, in the male group, two observers were red-green colour-blind. The effect of fatigue was considered to be a confounding factor. To allow for the influence of observer fatigue the order of the methods by which each observer made his measurements was varied by a regular system with each successive blood sample, so that all methods were used with equal frequency in all positions in the order of use.

Sadly, a full analysis of the results of the above experiment never seems to have been published. A tantalising glimpse of what might have been done is provided in the opening sentence of the authors' discussion:

> The experiment was so planned, with the help of R.A. Fisher that the various factors concerned in producing the over-all variability of the results of haemoglobin estimation by several observers could be assessed separately.

Similar measurement problems are faced in the behavioural and social sciences. These disciplines use a different terminology, however, and one that will seem very strange to anyone who has not come across it before. Instead of reproducibility they use *generalizability* and they embed their discussions of this issue within the realm of *generalizability theory* (Cronbach *et al.*, 1963; Gleser *et al.*, 1965; Brennan, 2001). And so, instead of being concerned with inter-laboratory precision studies they are concerned with the design and analysis of generalizability studies. The aims of a generalizability study, however, are very close to the above laboratory experiment to measure blood haemoglobin. In a generalizability study we set out systematically to investigate the sources of variation in social or psychological measurements. A measurement of the severity of depression, for example, is assumed to be a sample observation from a universe of permissible observations, characterised by one or more facets. For a given generalizability study we define a universe of admissible observations by listing the measurement conditions for each of the facets of interest. A facet (or experimental factor) can include, for example, alternative measurement instruments or psychometric test forms, different occasions, different laboratories or mental health clinics, or different scientists or clinicians within the laboratories or clinics, respectively. The different levels of a facet (such as alternative clinicians or mental health clinics) are called measurement or test conditions.

Consider, for example, the experiment on haemoglobin measurement described above. Here MacFarlane *et al.* (1948) considered five facets: type of measuring instrument (16 levels), sex of observer (2 levels), training of observer (2 levels), side of dominant eye (2 levels) and, finally, abnormalities in colour vision (2 levels). They also realised that there might be an effect of fatigue but, in their experiment, the order of making the measurements was varied to allow for this instead of it being explicitly recognised as a facet of observation. These authors did not, of course, use the terminology of the generalizability theorists, but their aims were the same.

A well known example of an early generalizability study is that reported by Gleser *et al.* (1978). This study examines the sources of variation in measures of psychological distress in disaster survivors. Each of 20 survivors provided data from two independent interviews. Each interview was then independently quantified by four trained raters using the Psychiatric Evaluation Form (PEF) developed by Spitzer and his colleagues (see Gleser *et al.*, 1978 for further details). There are, therefore, two facets: Interviewer (2 levels) and Rater (4 levels). A simpler and much more frequently used design would be the simple so-called inter-rater reliability study in which a behavioural characteristic of each of a sample of subjects is independently rated by each of a fixed panel of K raters. This design involves only a single facet: raters (with K levels).

But what is the universe of permissible observations? A test score or measurement on which a decision is to be based is one of many measurements that would adequately serve the same purpose.

> The decision maker is almost never interested in the response given to particular stimulus objects or questions, to the particular tester at the particular moment of testing. Some, at least, of these conditions of measurement could be altered without making the score any less acceptable to the decision maker.
>
> (Cronbach *et al.*, 1972)

An analytical chemist, for example, changes her batches of reagents, may move from one piece of equipment to another, or even change laboratories, without threatening the validity of the resulting measurement.

> That is to say, there is a universe of observations, any one of which would have yielded a usable basis for the decision. The ideal datum on which to base the decision would be something like the person's mean score over all acceptable observations, which we shall call his 'universe score'. The investigator uses the observed score or some function of it as if it were the universe score. That is, he generalizes from sample to universe. The question of 'reliability' thus resolves into a question of accuracy of generalization, or generalizability.
>
> (Cronbach *et al.*, 1972, page 15)

In any given investigation in which a measurement is used, however, the decision maker may choose to restrict the universe of admissible observations. He may, for example, decide that all subjects in a psychiatric research study should be interviewed and rated by a particularly well trained and experienced clinician. Following the haemoglobin study similar to that of MacFarlane and colleagues, subsequent investigators might choose to use a single analytical method under the control of trained observers known not to be red-green colour blind. This would still allow for variability in the observations due to the sex of the observer and the side of his or her dominant eye (and, possibly, fatigue). Clearly the reproducibility (generalizability) of the results in this revised study would be better than that in the original experiment. Moreover, we could use the results of the original study to estimate the variance components corresponding to each of the facets of the observations (and their interactions) and then use these estimates to infer the reproducibility in the simpler 'decision study'. Reproducibility or generalizability under a given set of permissible measurement conditions is 'simply' inferred from the appropriate combination of selected variance components estimated from the original comprehensive generalizability study.

1.7 More on the coefficient of reliability

Returning to Equation (1.1):

$$X_{ij} = \tau_i + e_{ij}$$

together with standard independence assumptions, it follows that, as in Equation (1.5),

$$\sigma_X^2 = \sigma_\tau^2 + \sigma_e^2$$

We define the coefficient of reliability (or reliability for short) for measurements of X as follows:

$$\kappa_X = \frac{\sigma_\tau^2}{\sigma_X^2} = \frac{\sigma_\tau^2}{\sigma_\tau^2 + \sigma_e^2} \tag{1.28}$$

An alternative expression (easily derived from (1.28)) is

$$\kappa_X = 1 - \frac{\sigma_e^2}{\sigma_X^2} = 1 - \frac{\sigma_e^2}{\sigma_\tau^2 + \sigma_e^2} \tag{1.29}$$

This expression leads immediately to the following expression for the standard error of measurement (σ_e – equivalent to the repeatability standard deviation) in terms of the reliability and the standard deviation of the measurements of X:

$$\sigma_e = \sigma_X \left(1 - \kappa_X\right)^{\frac{1}{2}} \tag{1.30}$$

It should be clear from Equation (1.29) that the reliability of X is not a fixed characteristic of the measurement process. Even if the standard error of measurement (σ_e) were fixed (and even this is frequently not the case) the reliability would vary with the heterogeneity of the population on which the measurements are being made. Increase σ_τ^2 and κ_X also increases. The reliability, κ_X, is simply the proportion of the variance of the observed measurements that is explained by variation in the true scores (that is the proportion that is not explained by measurement error). If the variance of the true scores were to drop towards zero then the reliability would also tend towards zero, even if the measurements were very precise (very low σ_e). This should always be borne in mind when using estimates of reliability derived from literature and attempting to use them for a population that may not display the same variability (that is, σ_τ^2) as the populations in the earlier studies. It is straightforward to use expressions such as (1.30) to calculate the reliability coefficient appropriate to the new population.

Yet another equivalent to Equation (1.28) is

$$\kappa_X = \frac{\sigma_{X\tau}}{\sigma_X^2} \tag{1.31}$$

where $\sigma_{X\tau}$ is the covariance of X and τ. It follows that

$$\text{Corr}(X, \tau) = \frac{\text{Cov}(X, \tau)}{\sqrt{\text{Var}(X)}\sqrt{\text{Var}(\tau)}} = \frac{\kappa_X \sigma_X^2}{\sigma_X \sigma_\tau}$$

$$= \kappa_X \sqrt{\frac{\sigma_X^2}{\sigma_\tau^2}} = \sqrt{\kappa_X} \tag{1.32}$$

If we have two parallel measurements (same τs, the same σ_e^2, and hence the same reliability, κ_X) on each subject ($X1$ and $X2$, say) then

$$\sigma_{X1}^2 = \sigma_{X2}^2 = \sigma_X^2$$

and

$$\text{Corr}(X1, X2) = \frac{\text{Cov}(X1, X2)}{\sqrt{\text{Var}(X1)}\sqrt{\text{Var}(X2)}}$$

$$= \frac{\sigma_\tau^2}{\sigma_X^2} = \kappa_X \tag{1.33}$$

This provides a basis for the estimation of reliability: measure a sample of subjects twice using parallel measures (repeats of the same measure, for example) and correlate the results. It is preferable, however, to use an estimate of the intra-class correlation rather than the more familiar Pearson product-moment correlation coefficient (see Section 2.2).

What if we were interested in calculating the reliability of the arithmetic mean (or sum) of $X1$ and $X2$? This is straightforward to derive but we cannot stress too highly that it is dependent on the assumption of the statistical independence of the measurement errors. If they are not independent (in many practical situations they are likely to be positively correlated, for example) then the gain in reliability will be an over-estimate. Consider the sum of $X1$ and $X2$:

$$X_{sum} = X1 + X2 = 2\tau + e1 + e2 \tag{1.34}$$

$$\text{Var}(2\tau) = 4\sigma_\tau^2 \tag{1.35}$$

$$\text{Var}(X_{sum}) = 4\sigma_\tau^2 + 2\sigma_e^2 \tag{1.36}$$

The reliability of the sum is then given by

$$\kappa_{sum} = \frac{4\sigma_\tau^2}{4\sigma_\tau^2 + 2\sigma_e^2} = \frac{2\sigma_\tau^2}{2\sigma_\tau^2 - \left(\sigma_X^2 - \sigma_\tau^2\right)}$$

$$= \frac{2\sigma_\tau^2}{\sigma_X^2 + \sigma_\tau^2} = \frac{2\sigma_\tau^2 / \sigma_X^2}{1 + \sigma_\tau^2 / \sigma_X^2}$$

$$= \frac{2\kappa_X}{1 + \kappa_X} \tag{1.37}$$

In general, if we would like to obtain the reliability of the sum of m parallel measurements (e.g. m replicates), then the following holds:

$$
\begin{aligned}
\kappa_{sum} &= \frac{m^2 \sigma_\tau^2}{m^2 \sigma_\tau^2 + m \sigma_e^2} \\
&= \frac{m^2 \kappa_X}{m^2 \kappa_X + m(1 - \kappa_X)} \\
&= \frac{m \kappa_X}{1 + (m - 1)\kappa_X}
\end{aligned}
\tag{1.38}
$$

Equation (1.38) is known as the Spearman-Brown prophesy formula.

Although we have not so far considered any measurements that cannot be regarded as quantitative, it is of interest to consider the characteristics of binary diagnostic tests in the identification or diagnosis of an underlying binary trait (see Chapter 5). Much of this work revolves around the estimation of agreement indices such as the well-known kappa coefficient of Cohen (1960) and/or in the estimation of a test's sensitivity and specificity. A test's sensitivity is the proportion of the trait positive (e.g. diseased) individuals who produce a positive test result. Specificity, on the other hand, is the proportion of the trait negative individuals who produce a negative test result. Here we will briefly look at the test's reliability (Kraemer, 1979). Suppose that X_{ik} is the kth replication of a binary measurement on the ith subject or individual. First,

$$
\Pr(X_{ik} = 1) = E_k(X_{ik}) = p_i
\tag{1.39}
$$

Kraemer (1979) refers to p_i as the ith subject's concensus score. From the properties of the binomial distribution:

$$
\mathrm{Var}(X_{ik}) = p_i(1 - p_i)
\tag{1.40}
$$

Assuming that all subjects are not all the same (that is, do not have the same consensus score) then it follows that this variance will vary from one subject to another. If we look at all of the subjects within the population, then

$$
E(p_i) = P
\tag{1.41}
$$

and

$$
\mathrm{Var}(p_i) = \sigma_p^2
\tag{1.42}
$$

Following the same rationale as for quantitative measurements, Kraemer (1979) defines reliability (the intraclass kappa coefficient) as the ratio of the concensus score variance to the observed score variance. That is,

$$
\kappa_X = \frac{\sigma_p^2}{P(1 - P)}
\tag{1.43}
$$

This idea has been generalized to multinomial nominal classifications by Kraemer (1979) and by Roberts and McNamee (1998), but we will not pursue that option here. The intraclass kappa coefficient is a particular example of an intraclass correlation, which is likely to be more familiar to investigators dealing with quantitative measurements (see Section 2.2), but many of the estimation procedures relevant to the latter can also be used for binary assessments (see Section 5.4, for example).

1.8 The rest of the book

The statistical evaluation of measurement errors involves studies of essentially two kinds. Although they are not mutually exclusive they are usually carried out as completely separate activities. The first is the reproducibility or generalizability study. Here the primary goal is usually the efficient estimation of variance components. These variance components may then be combined to produce various reliability coefficients and these may, in fact, be the main focus of the investigator. The second type of study is the method comparison study. Here we wish to investigate the covariation of measurements produced by two, three or more methods of measurement in view of estimation and comparison of their precisions. Scale conversion (relative calibration in the sense of estimating the αs and βs of a model such as (1.12)) may also be of interest. Again, it is possible to estimate reliability coefficients using the results of such a study. Method comparison studies can be carried out using either quantitative or binary measurements. A slightly different approach to method comparison (which appears to be particularly popular in clinical disciplines) is to evaluate the concordance or agreement between, say, any two given measurement techniques.

In Chapter 2 we have a detailed look at the sources of variability in quantitative measurements. In essence, the Chapter covers most of the material needed to evaluate the results of a reliability or generalizability study. Although there will be many references to reliability coefficients, most of the emphasis will be on the estimation of components of variance. Similar material for binary assessments will be found in Chapter 5. The major focus of this text, however, will be on the design and analysis of method comparison studies. Chapter 3 covers the situation when we have each of several individuals measured once using two measurement methods or instruments (that is, no replication). Although these designs are by far the most popular they have many disadvantages and these will be explored in detail. Chapter 4 covers the design and analysis of method comparison studies in which more information has been gathered. This could involve replication of the measurements taken by one or more of the methods, collection of data on so-called instrumental variables, or the use of more than two different measurement methods. This chapter is the core chapter of the book. Chapter 5 covers methods for binary assessments, including the assessment of agreement, reliability or generalizability, but, again, focussing on method comparisons. The ideas are illustrated throughout by reference to examples and greater emphasis is placed on this approach in the later chapters covering the more complex material (Chapters 4 and 5, in particular).

2

Sources of variation

The aim of this chapter is to familiarize the reader with the formal analysis of the variability of measurements through the estimation of variance components. We start with the estimation of the variance of a set of replicated measurements and then move on to one-way designs (in which each of a sample of subjects provides replicated measurements) and two-way inter-rater reliability studies (in which each subject is assessed by several raters). Finally, we consider examples in which there are missing observations, either by accident or by design. The main emphasis is on the use of traditional moments (ANOVA) estimators and methods of inferring their sampling characteristics based on Normality assumptions. The chapter also includes a discusssion of likelihood-based estimation procedures and the use of computationally-intensive methods such as the jack-knife and bootstrap.

2.1 A simple replication study

In Chapter 1 the concept of the precision of a measuring instrument was introduced through the example of a cook making repeated weighings of a single object using kitchen scales. A slightly more complicated experiment could have involved repeated weighings of each of a random sample of objects from the cook's kitchen. Table 2.1 presents the results of an analogous precision study using a simple map measurer. Each of ten map distances is measured five times, yielding a total of 50 measurements. Table 2.1 also gives the mean for each set of five measurements together with their variance. On the assumption that the precision of the map measurer is independent of the true length of the route being measured, this common precision can be estimated from the mean of the ten within-route error variances. The mean value is 0.413 with a standard error of 0.072. One can check whether the error variances are indeed homogeneous using one of several commonly used significance tests (see, for example, Caulcott & Boddy, 1983). In this particular example, the only likely source of lack of homogeneity is the dependence of the within-route variance (or standard deviation) on its mean.

The above estimate of the within-route error variance could have also been obtained through the use of a one-way analysis of variance (ANOVA). It is assumed that the reader is familiar with this technique (see, for example, Armitage, Berry & Matthews, 2002), and only a brief outline to introduce the necessary notation will be given here. Let the jth measurement

Table 2.1 Measurements of map distances (arbitrary units).

Route	Replicate measurements	Mean	Variance
1	48.0, 47.5, 48.0, 49.5, 48.0	48.2	0.575
2	32.0, 32.5, 33.5, 31.5, 32.5	32.4	0.550
3	12.5, 14.0, 14.0, 14.5, 14.5	13.9	0.675
4	29.5, 30.0, 29.0, 29.0, 30.0	29.5	0.250
5	50.0, 50.0, 49.5, 49.5, 50.5	49.9	0.175
6	55.5, 55.5, 56.0, 56.0, 56.5	55.9	0.175
7	22.0, 23.5, 23.0, 24.0, 22.0	22.9	0.800
8	50.5, 50.0, 51.5, 51.0, 50.0	50.6	0.425
9	26.5, 26.0, 26.0, 25.5, 25.5	25.9	0.175
10	24.0, 24.0, 24.5, 23.0, 23.5	23.8	0.325
		Mean:	0.413
		s.d.:	0.228

of the ith route be X_{ij} ($i = 1, 2, \ldots, 10; j = 1, 2, \ldots, 5$). Note that there is no correspondence between the jth measurement of any one route with the jth measurement of another. The repeated measurements are nested within routes and a better representation of the measurements might be $X_{j(i)}$ to indicate the jth replication within route i. Where there is no risk of confusion, however, the simpler terminology will be used. Another way of noting this structure is that the j replications within the ith route are interchangeable (their order of presentation is irrelevant). The mean of the five measurements for route i will be represented by $\bar{X}_i$, and the overall mean for the 50 measurements by $\bar{X}$.

In the more general situation where there are n_i measurements made on the ith subject (where $i = 1, 2, \ldots, I$ and $\Sigma_i n_i = N$) then

$$\bar{X}_i = \frac{1}{n_i} \sum_{j=1}^{n_i} X_{ij} \tag{2.1}$$

and

$$\bar{X} = \frac{1}{N} \sum_{i=1}^{I} \sum_{j=1}^{n_i} X_{ij} = \frac{1}{N} \sum_{i=1}^{I} \bar{X}_i \tag{2.2}$$

In the present example $I = 10$, $n_i = 5$ for all I, and $N = 50$. The total sum of squared deviations about the overall mean is given by

$$TSS = \sum_{i=1}^{I} \sum_{j=1}^{n_i} \left(X_{ij} - \bar{X} \right)^2 \tag{2.3}$$

The between-subjects (between-routes) sum of squares is given by

$$BSS = \sum_{i=1}^{I} \sum_{j=1}^{n_i} \left(\bar{X}_i - \bar{X} \right)^2 = n_i \sum_{i=1}^{I} \left(\bar{X}_i - \bar{X} \right)^2 \tag{2.4}$$

Table 2.2 ANOVA for the map distances in Table 2.1.

Source of variation	Degrees of freedom	Sum of squares	Mean square
Route	9	9562.000	1062.444
Error	40	16.500	0.4125

Finally, the within-subjects (within-routes) sum of squares is given by

$$WSS = \sum_{i=1}^{I} \sum_{j=1}^{n_i} \left(\overline{X}_{ij} - \overline{X}_i \right)^2 = \sum_{i=1}^{I} (n_i - 1)s_i^2 \tag{2.5}$$

where s_i^2 is the within-subject error variance for subject i.

The between-subjects mean square and the within-subjects mean square will be denoted by s_B^2 and s_W^2, respectively. These are calculated from

$$s_B^2 = \frac{BSS}{(I-1)} \tag{2.6}$$

and

$$s_W^2 = \frac{WSS}{(N-I)} \tag{2.7}$$

An analysis of variance for the map distances given in Table 2.1 is provided in Table 2.2. It should be clear from these data that s_W^2 is equivalent to the mean of the error variances given in Table 2.1.

2.2 Intra-class correlation

Looking at the map distance data in Table 2.1, how might we calculate the correlation between replicate measurements? When we calculate a Pearson product moment correlation coefficient between two measurements, X and Y, say, the X and Y are ordered (i.e. they are not exchangeable). The replicates in Table 2.1, however, are not in a particular order. In the case of two replicates per subject (or per route) the intra-class correlation is estimated from the product-moment correlation between pairs of measurements in which each pair enters the calculation twice (i.e. as $[X_{i1}, X_{i2}]$ and again as $[X_{i2}, X_{i1}]$). In general, if there are n_i measurements made on subject i there will be $n_i(n_i - 1)$ pairs of measurements for this one subject, each measurement being first in association with each of the other $(n_i - 1)$ measurements second. For each of the routes in Table 2.1, for example, there are 20 pairs of observations (i.e. 5×4) entering the calculation for each route (200 pairs in all).

In practice, one usually computes the intra-class correlation using a different procedure to that described above. As in the previous section, let the jth measurement on subject i be represented by X_{ij} ($i = 1, 2, ..., I; j = 1, 2, ..., J$). In the above correlation calculation each measurement on subject i will appear $n_i - 1$ times and so the mean of each of the variates being correlated (se X and Y) will be given by

$$\bar{X} = \bar{Y} = \frac{1}{N^*} \sum_{i=1}^{I} \left\{ (n_i - 1) \sum_{j=1}^{n_i} X_{ij} \right\} \tag{2.8}$$

where $N^* = \sum_{i=1}^{I} n_i(n_i - 1)$. Similarly,

$$\text{Var}(X) = \text{Var}(Y) = \frac{1}{(N^* - 1)} \sum_{i=1}^{I} \left\{ (n_i - 1) \sum_{j=1}^{n_i} \left(X_{ij} - \bar{X} \right)^2 \right\} \tag{2.9}$$

and

$$\text{Cov}(X, Y) = \frac{1}{(N^* - 1)} \sum_{i=1}^{I} \sum_{j=1}^{n_i} \sum_{h=1}^{n_i} \left(X_{ij} - \bar{X} \right) \left(X_{ih} - \bar{X} \right) \tag{2.10}$$

where the latter only applies to pairs of measurements for which $j \neq h$. The required Pearson product-moment correlation coefficient is given by

$$r_i = \frac{\displaystyle\sum_{i=1}^{I} \sum_{j=1}^{n_i} \sum_{h=1, h \neq j}^{n_i} \left(X_{ij} - \bar{X} \right) \left(X_{ih} - \bar{X} \right)}{\displaystyle\sum_{i=1}^{I} (n_i - 1) \sum_{j=1}^{n_i} \left(X_{ij} - \bar{X} \right)^2} \tag{2.11}$$

But,

$$\sum_{i=1}^{I} \sum_{j=1}^{n_i} \sum_{h=1, h \neq j}^{n_i} \left(X_{ij} - \bar{X} \right) \left(X_{ih} - \bar{X} \right) = \sum_{i=1}^{I} \sum_{j=1}^{n_i} \sum_{h=1}^{n_i} \left(X_{ij} - \bar{X} \right) \left(X_{ih} - \bar{X} \right)$$
$$- \sum_{i=1}^{I} \left(X_{ij} - \bar{X} \right)^2 \tag{2.12}$$

where, in this case, the summation over j and h now extends over all possible pairs, including $h = j$. Furthermore,

$$\sum_{i=1}^{I} \sum_{j=1}^{n_i} \sum_{h=1, h \neq j}^{n_i} \left(X_{ij} - \bar{X} \right) \left(X_{ih} - \bar{X} \right) = \sum_{i} n_i^2 \left(\bar{X}_i - \bar{X} \right)^2 \tag{2.13}$$

where, as before, $\bar{X}_i$ is the mean value for the measurements on the ith subject. Therefore

$$r_i = \frac{\displaystyle\sum_{i=1}^{I} n_i^2 \left(\bar{X}_i - \bar{X} \right)^2 - \sum_{i=1}^{I} \sum_{j=1}^{n_i} \left(X_{ij} - \bar{X} \right)^2}{\displaystyle\sum_{i=1}^{I} (n_i - 1) \sum_{j=1}^{n_i} \left(X_{ij} - \bar{X} \right)^2} \tag{2.14}$$

In the case where $n_i = n$ for all subjects, then

$$r_i = \frac{n^2 \sum\limits_{i=1}^{I} \left(\overline{X}_i - \overline{X}\right)^2 - \sum\limits_{i=1}^{I} \sum\limits_{j=1}^{n_i} \left(X_{ij} - \overline{X}\right)^2}{(n-1)\sum\limits_{i=1}^{I} \sum\limits_{j=1}^{n_i} \left(X_{ij} - \overline{X}\right)^2} \tag{2.15}$$

Substituting in this expression the various terms obtained from an analysis of variance table, it follows that

$$r_i = \frac{n \times BSS - TSS}{(n-1)TSS} \tag{2.16}$$

and since $TSS = BSS + WSS$, it follows that

$$\begin{aligned} r_i &= \frac{n \times BSS - (BSS + WSS)}{(n-1)(BSS + WSS)} \\ &= \frac{(n-1)(I-1)s_B^2 - I(n-1)s_W^2}{(n-1)\left[(I-1)s_B^2 - I(n-1)s_W^2\right]} \end{aligned}$$

simplifying to

$$r_i = \frac{(I-1)s_B^2 - Is_W^2}{(I-1)s_B^2 - I(n-1)s_W^2} \tag{2.17}$$

Further details can be found in Kendall (1943) or in Snedecor and Cochran (1967). For the data in Table 2.1, formula (2.17) gives an intra-class correlation of 0.998.

2.3 Expected mean squares

Consider the specification of a measurement model for the map distances in Table 2.1. One possibility is

$$X_{ij} = \tau_i + e_{ij} \tag{2.18}$$

This is equivalent to the model described in Equation (1.1). As in Equations (1.2) and (1.3) the expected values of the measurements are given by

$$E_j(X_{ij}) = \tau_i \tag{2.19}$$

and

$$E_{ij}\left(X_{ij}\right) = E_i\left(\tau_i\right) = \mu \tag{2.20}$$

τ_i is a random variable with mean μ and variance specified to be σ_τ^2. Similarly, e_{ij} is another random variable with mean zero and variance σ_e^2. It is assumed that the errors are uncorrelated with each other or with the other components of the model. The τ_i are also assumed to be uncorrelated. From Equation (2.18) and these assumptions it follows that

$$\text{Var}(X_{ij}) = \sigma_\tau^2 + \sigma_e^2 \tag{2.21}$$

$$\text{Cov}(X_{ij}, X_{ih}) = \sigma_\tau^2, (h \neq j) \tag{2.22}$$

and

$$\text{Cov}(X_{ij}, X_{kj}) = 0, (i \neq k) \tag{2.23}$$

Returning to the analysis of variance described in Section 2.1 and letting $n_i = n$ for all i, from Equations (2.4) and (2.6) it follows that

$$E\left(s_B^2\right) = \frac{n}{I-1} E\left[\sum_{i=1}^{I}\left(\overline{X}_i - \overline{X}\right)^2\right] = \frac{nI}{(I-1)} E\left(\overline{X}_i - \overline{X}\right)^2$$

$$= \frac{nI}{(I-1)}\left[E\left(\tau_i - \overline{\tau}\right)^2 + E\left(\overline{e}_i - \overline{e}\right)^2\right] \tag{2.24}$$

where $\overline{\tau}$ is the mean of the τ_i, $\overline{e}_i$ is the mean of the errors for subject i and $\overline{e}$ is the mean of all of the measurement errors. Therefore

$$E\left(s_B^2\right) = \frac{nI}{(I-1)}\left[\sigma_\tau^2 - \frac{\sigma_\tau^2}{I} + \frac{\sigma_e^2}{n} - \frac{\sigma_e^2}{nI}\right]$$

$$= n\sigma_\tau^2 + \sigma_e^2 \tag{2.25}$$

Similarly, from Equations (2.5) and (2.7)

$$E\left(s_W^2\right) = \frac{I}{N-I} E\left[\sum_{j=1}^{n}\left(X_{ij} - \overline{X}_i\right)^2\right]$$

$$= \frac{nI}{N-I} E\left(e_{ij} - \overline{e}_i\right)^2 = \frac{nI}{nI-I}\left[\sigma_e^2 - \frac{\sigma_e^2}{n}\right]$$

$$= \frac{nI}{I(n-1)}\left[\frac{n-1}{n}\right]\sigma_e^2 = \sigma_e^2 \tag{2.26}$$

The within-subjects mean square is therefore an unbiased estimate of the variance of the measurement errors. Similarly, the between-subjects variance, σ_τ^2, can be estimated from

$$\sigma_\tau^2 = \frac{s_B^2 - s_W^2}{n} \tag{2.27}$$

Finally, returning to the measurement model described in Equation (2.18), one can define the intra-class correlation by

$$\rho_i = \frac{\text{Cov}\left(X_{ij}, X_{ih}\right)}{\sqrt{\text{Var}\left(X_{ij}\right)\text{Var}\left(X_{ih}\right)}}, (j \neq h) = \frac{\text{Cov}\left(X_{ij}, X_{ih}\right)}{\text{Var}\left(X_{ij}\right)} = \frac{\sigma_\tau^2}{\sigma_\tau^2 + \sigma_e^2} \tag{2.28}$$

Equation (2.28) is also the definition of the measurements' reliability (see Section 1.7). One can substitute the estimates of σ_τ^2 and σ_e^2 into Equation (2.28) to obtain an estimate of this reliability as

$$\hat{\rho}_i = \frac{(1/n)\left[s_B^2 - s_W^2\right]}{(1/n)\left[s_B^2 - s_W^2\right] + s_W^2} = \frac{s_B^2 - s_W^2}{s_B^2 + (n-1)s_W^2} \tag{2.29}$$

For the data in Table 2.1, $\hat{\rho}_i = 0.998$. Note that Equations (2.17) and (2.29) are not equivalent, but for most practical purposes the difference is trivial.

2.4 Maximum likelihood estimators

Equations (2.21) to (2.23) define an $N \times N$ dispersion matrix V. If it is assumed that both the τ_i and the e_i are normally distributed, then the likelihood for the N measurements is given by

$$L = (2\pi)^{-\frac{1}{2}nI}|V|^{-\frac{1}{2}}\exp\{-\tfrac{1}{2}(X - \mu)'V^{-1}(X - \mu)\} \tag{2.30}$$

where the elements of X are the measurements, X_{ij}, arranged as an $N \times 1$ column vector and μ is an $N \times 1$ vector of μs (all identical). Equating to zero the differential of log L with respect to μ, σ_τ^2 and σ_e^2 provides the following solutions

$$\hat{\mu} = \overline{X} \tag{2.31}$$

$$\hat{\sigma}_e^2 = s_W^2 \tag{2.32}$$

$$\hat{\sigma}_\tau^2 = \left(\frac{I-1}{I}s_B^2 - s_W^2\right)\Big/n \tag{2.33}$$

Note that the ML estimator for σ_τ^2 is not the same as that given by Equation (2.27) and it is therefore biased. Substituting the solutions in (2.32) and (2.33) into Equation (2.28) gives

$$\hat{\rho}_i = \frac{\left\{[(I-1)/I]s_B^2 - s_W^2\right\}/n}{\left\{[(I-1)/I]s_B^2 - s_W^2\right\}/n + s_W^2}$$

$$= \frac{(I-1)s_B^2 - Is_W^2}{(I-1)s_B^2 + (n-1)Is_W^2} \tag{2.34}$$

This is identical to the expression in Equation (2.17).

A variant of ML estimation is to use restricted or residual maximum likelihood (REML) estimators. These are described in Searle (1987), Searle *et al.* (1992), Robinson (1987) and Harville (1977). The general techniques of REML estimation were developed by Patterson and Thompson (1971; 1975). An early example of its use for the analysis of the behaviour of measuring instruments is provided by Russell and Bradley (1958).

The technical details of REML estimation are beyond the scope of this book. (For details of both ML and REML estimation, the reader is referred to Searle *et al.* (1992)) but essentially the method involves maximizing the joint likelihood of a set of orthogonal contrasts of the measurements with all of the contrasts having zero expectation (i.e. error contrasts). In the case of balanced data (when $n_i = n$ for all i in the one-way ANOVA models, for example) REML estimators are identical to those obtained by equating mean squares with their expected values. One exception is when the estimates obtained using expected values are negative. Both ML and REML estimators are constrained to be positive or zero. Readers who are unfamiliar with the idea of linear contrasts and also orthogonality are referred to Hand and Taylor (1987) or Searle (1987) for a description of their properties. In the case of the model in (2.18) maximizing the likelihood of a set of error contrasts is equivalent to maximizing that part of the full likelihood that is invariant to the value of μ.

One advantage of ML and REML estimation is that asymptotic standard errors for parameter estimates can easily be obtained from the second derivatives of the likelihood equations and are usually produced by default by the appropriate software packages (See Section 2.6 for more details).

2.5 An inter-rater reliability study

Let's start by having a look at the string length data in Table 1.2. In Table 2.3 I have rearranged the data in a form that is suitable for analysis using a two-way analysis of variance (ANOVA). We assume that we no longer have access to the standard (S) obtained by using a rule. The last column (mean3) is the mean of the three raters' guesses for the given piece of string. For the present, however, we ignore this last column. Table 2.4 provides the results of a two-way ANOVA (guess by string and rater) for these data. Note that without replication by each of the raters there is no way of distinguishing the string by rater interaction and the measurement error (i.e. lack of individual rater precision) in this analysis. They are said to be aliased or confounded – hence the use of the term 'residual' rather than 'error' in Table 2.3. Note also that the analysis is based on the assumption that the variance of each of the raters' residuals is the same as that of the others. A test of the quality of rater effects can be constructed from the ratio of the rater and residual mean squares ($F = 0.567/0.112 = 5.05$ with 2 and 28 d.f.: $P = 0.014$). So far we have learnt that the raters differ and that there are residual measurement 'errors' after allowing for the effects of rater, but not much else. Where do we go from here? We can consider the three raters as a random sample from a much larger population of raters about which we might wish to make inferences, or we can regard Graham, Brian and Andrew as the only raters of interest and treat them as fixed. We start with the raters being treated as a random sample.

2.5.1 Raters random

First, let's consider an appropriate statistical model. The measurement model for the two-way ANOVA has the form

$$X_{ij} = \tau_i + \alpha_j + \gamma_{ij} + e_{ij} \tag{2.35}$$

where X_{ij} is the guess of the length of the ith piece of string by the jth rater, τ_i is the effect of the ith piece of string (with mean μ and variance σ_τ^2), α_j is the effect of the jth rater (with mean 0 and variance σ_α^2), γ_{ij} is the interaction between string i and rater j (mean 0 and

Table 2.3 Estimates of the lengths of 15 pieces of string (rearranged data from Table 1.2).

String	Rater*	Guess	Mean3	String	Rater*	Guess	Mean3
1	1	5.0	5.3	8	3	5.5	4.8
1	2	4.8	5.3	9	1	2.4	2.2
1	3	6.0	5.3	9	2	2.0	2.2
2	1	3.2	3.3	9	3	2.1	2.2
2	2	3.1	3.3	10	1	5.2	5.5
2	3	3.5	3.3	10	2	5.3	5.5
3	1	3.6	4.0	10	3	6.0	5.5
3	2	3.8	4.0	11	1	1.2	1.2
3	3	4.5	4.0	11	2	1.1	1.2
4	1	4.5	4.3	11	3	1.2	1.2
4	2	4.1	4.3	12	1	1.8	1.7
4	3	4.3	4.3	12	2	1.6	1.7
5	1	4.0	4.7	12	3	1.8	1.7
5	2	5.2	4.7	13	1	3.4	3.8
5	3	5.0	4.7	13	2	4.1	3.8
6	1	2.5	2.6	13	3	3.9	3.8
6	2	2.8	2.6	14	1	6.0	6.3
6	3	2.6	2.6	14	2	6.3	6.3
7	1	1.7	1.6	14	3	6.5	6.3
7	2	1.4	1.6	15	1	2.2	1.9
7	3	1.6	1.6	15	2	1.6	1.9
8	1	4.8	4.8	15	3	2.0	1.9
8	2	4.2	4.8				

*Raters: 1 – Graham, 2 – Brian, 3 – Andrew.

Table 2.4 Two-way ANOVA for the string data.

Source of variation	Sum of squares	Degrees of freedom	Mean square
String	109.010	14	7.786
Rater	1.134	2	0.567
Residual error	3.146	28	0.112
Total	113.29	44	

variance σ_γ^2), and, finally, e_{ij} is the corresponding measurement error (mean 0 and variance σ_e^2). Without replication γ_{ij} and e_{ij} are indistinguishable and the model becomes

$$X_{ij} = \tau_i + \alpha_j + r_{ij} \tag{2.36}$$

where $r_{ij} = \gamma_{ij} + e_{ij}$. If we assume that γ_{ij} and e_{ij} are uncorrelated, then

$$\sigma_r^2 = \sigma_\gamma^2 + \sigma_e^2 \tag{2.37}$$

If γ_{ij} and e_{ij} are also uncorrelated with the subject and rater effects, and the last two are also uncorrelated with each other, then

$$\mathrm{Var}(X_{ij}) = \sigma_\tau^2 + \sigma_\alpha^2 + \sigma_\gamma^2 + \sigma_e^2$$
$$= \sigma_\tau^2 + \sigma_\alpha^2 + \sigma_r^2 \tag{2.38}$$

Equations (2.35) and (2.36) describe two-way random effects models. The reliability of the measurement of a randomly-selected subject (here, piece of string) by a randomly selected rater (both subjects and raters being selected from potentially infinite populations of subjects and raters, respectively) is an intra-class correlation, defined by

$$\rho_1 = \frac{\sigma_\tau^2}{\sigma_\tau^2 + \sigma_\alpha^2 + \sigma_e^2} \tag{2.39}$$

If this reliability is not sufficiently high, then we can replicate the measurements, and the reliability of the mean of the assessments of m independent raters on a given subject is given by

$$\rho_m = \frac{\sigma_\tau^2}{\sigma_\tau^2 + \left(\sigma_\alpha^2 + \sigma_e^2 \right)/m} \tag{2.40}$$

If, however, the m measurements are all independent replicates from only one randomly selected rater, then

$$\rho_{m(1)} = \frac{\sigma_\tau^2}{\sigma_\tau^2 + \sigma_\alpha^2 + \left(\sigma_e^2/m \right)} \tag{2.41}$$

The reliability in (2.40) can, alternatively, be calculated using the Spearman-Brown formula (see Section 1.7), but that in (2.41) cannot.

We can use arguments similar to those described in Section 2.3 to derive the expected values of the mean squares from the two-way ANOVA in terms of these unknown variance components. These are shown in Table 2.5(a). We can now equate observed mean squares with their expected values and solve for the unknown variance components to obtain the following estimates.

$$\hat{\sigma}_\tau^2 = \frac{SMS - EMS}{J} \tag{2.42}$$

$$\hat{\sigma}_\alpha^2 = \frac{RMS - EMS}{I} \tag{2.43}$$

$$\hat{\sigma}_e^2 = EMS \tag{2.44}$$

Table 2.5 Expected mean squares for a two-way ANOVA.

(a) Without replication: raters random (I subjects each assessed by J raters)

Source of variation	Abbreviation	d.f.	Expected mean square
Subject (e.g. string)	SMS	$I - 1$	$J\sigma_\tau^2 + \sigma_e^2$
Rater	RMS	$J - 1$	$I\sigma_\alpha^2 + \sigma_e^2$
Residual error	EMS	$(I - 1)(J - 1)$	σ_e^2

(b) With replication: raters random (I subjects each assessed by J raters; n observations per cell)

Source of variation	Abbreviation	d.f.	Expected mean square
Subject (e.g. string)	SMS	$I - 1$	$Jn\sigma_\tau^2 + n\sigma_\gamma^2 + \sigma_e^2$
Rater	RMS	$J - 1$	$In\sigma_\alpha^2 + n\sigma_\gamma^2 + \sigma_e^2$
Subject × rater	SRMS	$(I - 1)(J - 1)$	$n\sigma_\gamma^2 + \sigma_e^2$
Error	EMS	$IJ(n - 1)$	σ_e^2

Substituting the estimates from (2.42), (2.43) and (2.44) into Equation (2.39) yields the following estimate for the intra-class correlation:

$$\hat{\rho}_1 = \frac{I(SMS - EMS)}{I \times SMS + J \times RMS + (IJ - I - J)EMS} \tag{2.45}$$

This estimator was first derived by Bartko (1966). The estimates of σ_τ^2, σ_α^2 and σ_e^2 obtained from the ANOVA table for the string data are 2.558, 0.030 and 0.112, respectively. $\hat{\rho}_1 = 0.947$.

2.5.2 Raters fixed

In this situation we do not regard Graham, Brian and Andrew as a sample from a much larger population of potential raters. Here, they are the only raters of interest. The rater effects are said to be fixed. A model such as that described by Equations (2.35) or (2.36), in which there is a mixture of random (e.g. string) and fixed (e.g. rater) effects is called a two-way mixed effects model. In both the random effects and the fixed effects model the interaction, γ_{ij}, is considered to be a random variable with mean 0 and variance σ_γ^2. The expected values of the variances of the random components for model (2.36) are as in Table 2.5(a) and the variance components are estimated by equating observed and expected values and solving for the unknown variance components. However, the expected mean square for the rater effects is now given by

$$\sigma_e^2 + \frac{I}{J-1} \sum_j^3 \alpha_j^2 \tag{2.46}$$

The reliability of a measurement on a single randomly selected subject (from the large population of subjects) made by a single rater randomly selected from our fixed panel of J raters is given by

$$\rho_{1*} = \frac{\sigma_\tau^2}{\sigma_\tau^2 + \frac{1}{J}\sum \alpha_j^2 + \sigma_e^2} \tag{2.47}$$

If, however, we are only interested in the performance of each of the raters separately (e.g. what is the reliability of Brian's assessments?) then

$$\rho_{1**} = \frac{\sigma_\tau^2}{\sigma_\tau^2 + \sigma_e^2} \tag{2.48}$$

For the string data, $\hat{\rho}_{1*} = 0.958$. In the situation where a given subject is independently assessed by m randomly selected raters from the panel of J raters, then the reliability of the mean of the m ratings for this subject is given by

$$\rho_{m*} = \frac{\sigma_\tau^2}{\sigma_\tau^2 + \left(\frac{1}{J}\sum \alpha_j^2 + \sigma_e^2\right)/m} \tag{2.49}$$

2.5.3 Making sense of interactions

Thinking of Graham, Brian and Andrew's attempt to guess the lengths of pieces of string, or, in the context of a chemistry laboratory, the comparison of the performance of different types of assay, it is clearly of interest to question whether there is more to rater or instrument biases than the main effects for raters (i.e. the αs in Equations (2.35) and (2.36), for example) picked up by the two-way ANOVA. In Sections 1.3 to 1.5 we discussed proportional or relative biases (represented by the βs in Equations (1.11) and (1.12), for example). Can we extract any information concerning these βs from a two-way ANOVA, with or without replications by the raters? That is, is there a way of estimating these βs from information contained in the subject by rater interactions? The answer is 'Yes', and, perhaps surprisingly, even in the case where there is no replication. My statement about the complete aliasing of interactions and pure measurement error in a two-way table from an inter-rater reliability study without replication is not strictly true. We can get more from Table 2.3, for example. Before continuing the analysis of the real data, however, we will illustrate how one might approach the analysis from simulated string data with duplicate guesses by each rater. We will then drop one of the duplicates and show that one can still investigate the effects of proportional biases in the absence of replication. Finally, we will return to look at further results for the real string data. The analyses involve some approximations and are therefore not the perfect solution to our problem. But they are relatively simple to carry out and naturally lead on to the more informative and appropriate measurement error models to be described in Chapter 4.

Consider the data in Table 2.6. This is a Monte Carlo simulated data set, which has properties similar to that of the string data in Tables 1.2 and 2.3. The data represent 25 simulated pieces of string (the truth, τ_i, normally distributed with mean 5.0 and standard deviation of 2.0). Each piece of string was then independently 'rated' twice by three separate raters. For the ith piece of string, Rater 1 produced ratings obeying the linear relationship $X_{i1} = 1.2 \times \tau_i + e_1$ where the errors (e_{i1}s) were normally distributed with mean zero and standard deviation 0.3 (i.e. variance 0.09). Similarly, for Rater 2, $X_{i1} = 1.0 \times \tau_i + e_{i2}$ and for Rater 3, $X_{i1} = 0.8 \times \tau_i + e_{i3}$, where the measurement errors for both raters had means of zero and standard deviation 0.3. In addition to these data, Table 2.6 also gives the mean of the six guesses for each piece of string (mean6). The mean of the first observations from each rater (mean3) was also calculated for each piece of string (not shown). All values were truncated to two decimal places.

Let's start with a two-way ANOVA. The results are shown in Table 2.7. Assuming that rater effects are random, it is straightforward to estimate the relevant variance components from the expected values of the mean squares given in Table 2.5(b). If we were to consider raters as fixed we would simply replace the $In\sigma_\alpha^2$ on the second row of this table by $(In/J)\Sigma_j\alpha_j^2$ (again assuming that $\Sigma_j\alpha_j = 0$). Calculation of reliabilities from the resulting variance components is also relatively straightforward. Instead of doing this, however, we will aim to model the γ_{ij}s (i.e. the interaction terms) in a way that is both informative and convincing from our *a priori* knowledge. Instead of assuming that raters have a constant bias relative to the others, we assume that it is proportional (relative). That is, we assume, for example that the bias is dependent upon the true value of the string's length (τ_i) in such away that it is relatively small for short pieces of string and larger for the longer ones. Using this argument, Healy (1989) suggests the following model:

$$X_{ijk} = \tau_i + \alpha_j + \beta_j\tau_i + e_{ijk} \tag{2.50}$$

Table 2.6 Simulated string data – with replication.

String	Rater 1		Rater 2		Rater 3		Mean6
1	5.62	5.59	4.41	4.41	4.36	3.89	4.71
2	8.78	8.91	6.53	6.89	5.96	5.01	7.02
3	5.91	6.02	4.88	4.98	4.13	3.99	4.98
4	6.17	5.07	4.49	4.67	4.27	4.08	4.79
5	3.52	3.71	3.13	2.91	2.66	2.32	3.04
6	11.40	11.05	9.25	8.81	7.17	7.44	9.18
7	2.47	1.56	1.98	1.24	1.60	1.74	1.76
8	4.82	5.35	4.59	3.94	3.69	4.03	4.40
9	4.82	4.71	4.51	3.93	3.48	2.99	4.07
10	10.45	10.50	8.96	8.54	7.26	6.96	8.78
11	6.20	5.45	4.88	5.02	4.14	4.12	4.97
12	6.86	6.24	5.36	5.16	4.11	4.35	5.35
13	6.61	7.10	6.18	5.33	4.93	4.99	5.86
14	2.00	1.50	1.76	1.24	1.34	0.62	1.41
15	9.19	8.76	7.22	7.41	6.30	5.99	7.48
16	5.94	5.52	4.65	4.55	4.36	4.02	4.84
17	6.88	7.14	4.92	5.51	3.80	4.56	5.47
18	5.47	5.33	4.87	4.35	3.24	3.71	4.49
19	3.72	3.64	2.76	3.04	1.80	2.23	2.87
20	3.91	3.93	2.96	3.62	2.92	2.38	3.29
21	6.87	7.65	6.49	6.50	5.02	4.91	6.24
22	6.32	6.11	5.48	5.43	4.92	4.42	5.45
23	7.38	7.95	6.36	6.54	5.63	5.12	6.49
24	9.41	9.02	7.74	7.76	5.97	5.67	7.60
25	5.44	6.21	4.82	5.43	4.13	3.55	4.93

Table 2.7 Two-way ANOVA for simulated string data (with replication).

(a) Initial ANOVA

Source of variation	SS	d.f.	MS
Model	656.07	74	
String	535.75	24	
Rater	100.00	2	50.00
String × rater	20.32	48	0.423
Error	7.76	75	0.103
Total	663.83	149	

(b) Model with proportional biases (sequential sums of squares)

Source of variation	SS	d.f.	MS
Model	656.07	74	
String	535.75	24	
Rater	100.00	2	50.00
Proportional biases	15.57	2	7.79
Residual interaction	4.65	46	0.101
Error	7.76	75	0.103
Total	663.83	149	

Note that the τ_i appear twice in this model and the model is, of course, equivalent to

$$X_{ijk} = \alpha_j + \beta_j^* \tau_i + e_{ijk} \qquad (2.51)$$

which, in turn, is the basic latent variable model described in Chapter 1 (see Equation (1.11)). Methods of fitting this latent variable model will be introduced in Chapter 4, but here we make a simple approximation by replacing the second τ_i in Equation (2.50) by an estimate based on the observed mean of the observations on the ith subject ($\overline{X}_i$). In the context of the simulated string data this is the mean of the six guesses for each piece of string (two replicates from each of three raters). This is the variable 'mean6' in the column to the right of Table 2.6. So, the revised model is now

$$X_{ijk} = \tau_i + \alpha_j + \beta_j \overline{X}_i + e_{ijk} \qquad (2.52)$$

This appears to have been first used by Mandel (1959) to analyse data from inter-laboratory precision studies and by Mosteller to analyse inter-rater reliability (unpublished work cited in Cochran, 1968). It is a model dealt with in some detail in books by Mandel (1991, 1995) and will be used again when we come to interpretation of the results of inter-laboratory precision studies (see Sections 4.9 and 4.11). The general problem of modelling the interaction between a qualitative factor (e.g. rater) and a quantitative one (e.g. string length) has a long history going back to work by Fisher and Mackenzie (1923). The latter authors were concerned with the yields from crops as a function of type of fertilizer (a qualitative factor) and the nitrogen content of the fertilizer (the quantitative factor). Williams (1952) provides a detailed discussion of the methods to solve the general problem, and Cochran (1943) is an early example of the exploration of interaction in terms of measurement error biases. Typically, the structure of the table of interactions (the estimates of the γ_{ij}s) is approached through a singular value decomposition (SVD). This is beyond the scope of the present text, but Mandel (1991, 1995) provides a relatively non-technical description of the use of the SVD in the context of a two-way table of measurements.

Having extracted a component of the interaction sum of squares due to proportional biases, one might ask what remains. One possibility, of course, is that it is simply the contribution to the random measurement errors (the e_{ij}s). Another possibility is a term that will change completely at random from one subject-rater combination to another, but will remain constant on replication of the individual measurements. Such a term is an example of a random matrix or interference effect as will be described in much more detail in Sections 3.9 and 4.10 (but see Equation 1.25).

Returning to the model in (2.52), Table 2.7(b) shows the results of fitting this model to the simulated string data. Note that we have partitioned the variation using sequential sums of squares (i.e. the effect of a variable in (2.52) is conditional on the effects to the left of it in the equation – or, in terms of the ANOVA table, conditional on the effects above it in the table). About 77% of the variation in the interactions is explained by the proportional biases (i.e. the βs). Estimates of the βs are given in Table 2.9(a). They are very close to their known values.

Now let's drop the second replicate for each of the raters in the simulated data set in Table 2.6. We now have only three measurements for each piece of string (i.e. one from each of the three raters). The corresponding ANOVA is given in Table 2.8(a). Note that we have fitted an interaction term which has led to a zero error sum of squares (the two sums of squares are aliased). We could have dropped the interaction from the model and the same sum of squares would then have appeared as that due to residual error. We now calculate the mean of each of these three measurements for each piece of string separately and again fit the model in Equation (2.52) to this reduced data set. The results are shown in Table 2.8(b). Note that the proportion of the sum of squares of the interactions explained by the proportional biases (i.e. 6.91/12.30) is about the same as the proportion of the sum of interaction and error sums of squares explained by proportional biases in Table 2.7 (i.e. 15.57/28.08). The main point of this analysis, however, is to demonstrate that these proportional biases can be estimated from a data set without replications. The required estimates are provided in Table 2.9(b). They are very similar to those obtained using the full data set but, understandably, they have larger standard errors.

Now let's have another look at the real string data (Table 2.3). Fitting the model in Equation (2.52) yields the estimates given in Table 2.10(a). It is worth noting that instead of

Table 2.8 Two-way ANOVA for simulated string data (without replications).

(a) Initial ANOVA

Source of variation	SS	d.f.	MS
Model	323.75	74	
String	263.32	24	
Rater	48.17	2	24.06
String × rater	12.30	48	0.256
Error	0	0	
Total	323.75	75	

(b) Model with proportional biases (sequential sums of squares)

Source of variation	SS	d.f.	MS
Model	323.75	74	
String	263.32	24	
Rater	48.13	2	24.06
Proportional biases	6.91	2	3.46
Residual interaction	5.38	46	0.117
Error	0	0	
Total	323.75	74	

Table 2.9 Estimates of proportional biases (subject to constraint $\Sigma_j \beta_j = 1$).

(a) With replications

	$\hat{\beta}_j$	se($\hat{\beta}_j$)
Rater 1	1.21	0.02
Rater 2	0.99	0.02
Rater 3	0.80	0.02

(b) Without replications

	$\hat{\beta}_j$	se($\hat{\beta}_j$)
Rater 1	1.20	0.03
Rater 2	0.99	0.03
Rater 3	0.81	0.03

Table 2.10 Estimates of proportional biases for real string data (subject to constraint $\Sigma_j \beta_j = 1$).

(a) Analysis of raw data

	$\hat{\beta}_j$	$se(\hat{\beta}_j)$
Rater 1	0.89	0.04
Rater 2	1.00	0.04
Rater 3	1.11	0.04

(b) Analysis of paired differences

	$\widehat{\beta_1 - \beta_2}$	$se(\widehat{\beta_1 - \beta_2})$
Pair 1	−0.10	0.07
Pair 2	−0.22	0.07

Pair 1 = Graham–Brian
Pair 2 = Graham–Andrew

fitting the model to the data from the raters directly, we could have fitted a similar model to the differences between the observations from different raters (see, for example, Cochran, 1943). Here we choose the differences between Graham and Brian (Pair 1) and those between Graham and Andrew (Pair 2). The estimates of the effects of string length on these differences (estimates of $\beta_1-\beta_2$ and $\beta_1-\beta_3$, respectively) are given in Table 2.10(b). They are entirely consistent with the earlier findings. Note that if we were to have measurements (without replication) from only two raters then this analysis is that based on the well-known Bland-Altman plot used in method comparison studies (see Section 3.2). However, in the case of only two raters the relative biases can only be estimated if we make *a priori* assumptions concerning the relative sizes of the measurement error variances for the two raters. The results presented in this section are based on assuming that they are equal.

It must be stressed, in any case, that the model in (2.52) is only an approximation to the one we are really interested in (that is, in Equations (2.50) and (2.51)). Methods for fitting models such as (2.51) will be discussed in detail in Chapters 3 and 4. As well being based on the explicit recognition that we are trying to model the effects of a latent variable (the unknown string length, for example) they will also allow for the fact that the measurement error variances of the raters might differ.

Before leaving this section we will briefly mention ML and REML estimation. Using these methods it is possible to get closed form estimates for variance components in the two-way mixed model (with or without the interaction), but we will not bother to present them here. For the two-way random effects model, however, there are no closed form estimates available (again, with or without interaction). For the random effects models we have to use iterative methods. Again, we will not get involved in the technical details (see Searle *et al.*, 1992).

2.6 Sampling distributions

The simplest, but not always feasible, way of investigating sampling distributions is to repeat the measurement study over and over again and explicitly examine how the required estimates vary from one repetition to another. Brennan (2001, Table 6.1) provides a convincing example. The variance estimates for each row of Table 2.1 (the variances of the within-route map distances) are another. Here, however, we develop inferential methods based on the results from a single study.

2.6.1 Distribution of an estimate of a variance from replicate measurements from a single subject

In Chapter 1 we illustrated the idea of variability of measurement errors using a very simple experiment in which a cook weighed a packet of food several times. The estimated variance (s^2) obtained from his first ten measurements was 250. If we assume that the measurement errors are independently and identically normally distributed with a mean of zero and variance σ^2, then it is well known that

$$\frac{\sum_i (X_i - \overline{X})^2}{\sigma^2}$$

is distributed as a χ^2-distrubution with $n - 1$ degrees of freedom (n being the number of replicated measurements) – see, for example, Armitage et al. (2002; pp147–53).

$$s^2 = \frac{\sum_i (X_i - \overline{X})^2}{n - 1} = \frac{\sigma^2}{n - 1} \frac{\sum_i (X_i - \overline{X})^2}{\sigma^2} = \frac{\sigma^2}{n - 1} \chi^2_{(n-1)} \tag{2.53}$$

The mean of a χ^2 with $n - 1$ degrees of freedom is $n - 1$, and its variance is $2(n - 1)$. It follows therefore that the expected value of s^2 is given by

$$E(s^2) = \frac{\sigma^2}{n - 1}(n - 1) = \sigma^2 \tag{2.54}$$

and its variance is given by

$$\text{Var}(s^2) = \frac{\sigma^4}{(n - 1)^2} 2(n - 1) = \frac{2\sigma^4}{n - 1} \tag{2.55}$$

From the definition of a variance, we have

$$\text{Var}(s^2) = E[s^2 - E(s^2)]^2 = E[(s^2)^2] - [E(s^2)]^2 \tag{2.56}$$

Equating (2.55) and (2.56) gives

$$E\left[(s^2)^2\right] = \left[E(s^2)^2\right]\left(1 + \frac{2}{n - 1}\right) \tag{2.57}$$

and therefore

$$\frac{E\left[(s^2)^2\right]}{(n - 1) + 2} = \frac{E\left[(s^2)\right]^2}{(n - 1)} \tag{2.58}$$

So, we could estimate the variance of s^2 by $s^4/(n - 1)$ – that is, by replacing σ in (2.55) by its estimated value, s, but an unbiased estimate of the variance of s^2 is given by

$s^4/[(n-1)+2]$. In our cook's data, s^2 was 250, and the variance of s^2 is therefore estimated by $250^2/11 = 5681$ (equivalent to a standard error of 75.4). The coefficient of variation of s^2 is given by its standard deviation divided by its expected value, that is, $\sqrt{2/(n-1)}$. Using the delta technique we can show that the variance of s is $\sigma^2/2(n-1)$ and the coefficient of variation of s is $1/\sqrt{[2(n-1)]}$. In designing our simple replication study, Healy (1989) argues that we should be aiming for a coefficient of variation for s of about 10%. To get this n needs to be about 50.

What about confidence intervals? If we are prepared to assume that the errors are normally distributed then

$$\chi^2_{(n-1),\alpha/2} < \frac{\sum_i \left(X_i - \overline{X}\right)^2}{\sigma^2} < \chi^2_{(n-1),1-\alpha/2} \tag{2.59}$$

where $\chi^2_{(n-1),p}$, for example, is the value of $\chi^2_{(n-1)}$ corresponding to a cumulative probability of p (i.e. $\text{Prob}(\chi^2_{(n-1)} < \chi^2_{(n-1),p}) = p$). From (2.59) we obtain

$$\frac{\sum_i\left(X_i - \overline{X}\right)^2}{\chi^2_{(n-1),1-\alpha/2}} < \sigma^2 < \frac{\sum_i\left(X_i - \overline{X}\right)^2}{\chi^2_{(n-1),\alpha/2}} \tag{2.60}$$

Using our cook's data and choosing $\alpha = 0.05$ (i.e. for a 95% two-sided confidence interval), we find that $\chi^2_{9,0.025} = 2.70$ and $\chi^2_{9,0.975} = 19.02$ and the sum of squares is 2000 (i.e. 9×250). Hence

$$\frac{2000}{19.02} < \sigma^2 < \frac{2000}{2.70}$$

The required 95% confidence interval is therefore (105, 741). One simply takes the square root of the lower and upper limits to get the corresponding interval for s: (10.2, 27.2). As one would expect from such a small sample these intervals are fairly wide. It is straightforward to use the jack-knife or bootstrapping (see, for example, Armitage *et al.*, 2002: pages 298–306) to investigate the sampling properties of $\hat{\sigma}^2$ if one is not convinced of the normality of the measurement errors. We discuss these methods in some detail in Section 2.7.

2.6.2 Variance components from the one-way ANOVA

Under normality, and on the assumption that the number of observations per subject is the same for subjects (i.e. the design is balanced), it is possible to show that the subject and error sums of squares are independently distributed (see, for example, Searle *et al.*, 1992). For a given mean square, MS,

$$\text{Var}(MS) = \frac{2[E(MS)]^2}{v} \tag{2.61}$$

where v is the degrees of freedom for the MS, and also that $(MS)^2/(v+2)$ is an unbiased estimator of $[E(MS)]^2/v$. It follows that

$$\text{Var}\left(\sigma_e^2\right) = \frac{2\sigma_e^4}{I(n-1)} \tag{2.62}$$

Similarly,

$$\mathrm{Var}(\sigma_\tau^2) = \mathrm{Var}\left(\frac{\left(s_B^2 - s_w^2\right)}{n}\right)$$

$$= \frac{2}{n^2}\left[\frac{\left(n\sigma_\tau^2 + \sigma_e^2\right)^2}{I - 1} + \frac{\sigma_e^4}{I(n - 1)}\right] \tag{2.63}$$

An unbiased estimator is

$$\frac{2}{n^2}\left[\frac{(n\hat\sigma_\tau^2 + \hat\sigma_e^2)^2}{I + 1} + \frac{\sigma_e^4}{I(n - 1) + 2}\right] \tag{2.64}$$

Under normality assumptions, the ML and REML estimator for $\mathrm{Var}(\hat\sigma_e^2)$ is the same as in Equation (2.61). That for $\hat\sigma_\tau^2$ is similar to Equation (2.62), but the term $(I - 1)$ (i.e. the denominator for the first part of the expression within the square brackets) is replaced by $I^2/(I - 1)$. Further details are provided by Searle *et al.* (1992). It is straightforward to derive confidence intervals for σ_e^2 using arguments similar to those in Section 2.6.1, but there is no exact confidence interval for σ_τ^2. However, we are not particularly interested in this parameter on its own, but we are much more likely to want a confidence interval for the reliability ratio, $\sigma_\tau^2/(\sigma_\tau^2 + \sigma_e^2)$. It is a bit more involved. We start by finding a confidence interval for the signal-to-noise ratio, $q = \sigma_\tau^2/\sigma_e^2$ and then move on to get the corresponding interval for $\rho_i = q/(q + 1)$, the reliability coefficient. First, however, we remind ourselves of the definition of an F_{ν_1,ν_2}-distribution. This is the probability distribution of $(\chi_{\nu_1}^2/\nu_1)/(\chi_{\nu_2}^2/\nu_2)$ where $\chi_{\nu_1}^2$ and $\chi_{\nu_2}^2$ are independent χ^2-variates with ν_1 and ν_2 degrees of freedom, respectively. From the distributional properties of the mean squares, we have

$$F_{\nu_1,\nu_2} = \frac{s_B^2/\sigma_B^2}{s_W^2/\sigma_W^2} = \frac{s_B^2}{s_W^2}\frac{\sigma_e^2}{\left(n\sigma_\tau^2 + \sigma_e^2\right)}$$

$$= F\frac{\sigma_e^2}{\left(n\sigma_\tau^2 + \sigma_e^2\right)} \tag{2.65}$$

where F is the observed F-ratio derived from the one-way ANOVA table given by $F = s_B^2/s_W^2$. This follows from $E(s_W^2) = \sigma_W^2 = \sigma_e^2$ and $E(s_B^2) = \sigma_B^2 = n\sigma_\tau^2 + \sigma_e^2$.
Now,

$$F_L < F\frac{\sigma_e^2}{\left(n\sigma_\tau^2 + \sigma_e^2\right)} < F_U \tag{2.66}$$

provides a confidence interval for F_{ν_1,ν_2}, where F_L and F_U are the lower and upper critical values of the corresponding F-distribution with the cumulative probabilities, $p = \alpha/2$ and $p = 1 - \alpha/2$, respectively. This can be rearranged to produce

$$\frac{F_L}{F} < \frac{\sigma_e^2}{\left(n\sigma_\tau^2 + \sigma_e^2\right)} < \frac{F_U}{F} \tag{2.67}$$

This is equivalent to

$$\frac{F}{F_U} < \frac{\left(n\sigma_\tau^2 + \sigma_e^2\right)}{\sigma_e^2} < \frac{F}{F_L} \tag{2.68}$$

and, therefore, to

$$\frac{F/F_U - 1}{n} < \frac{\sigma_\tau^2}{\sigma_e^2} < \frac{F/F_L - 1}{n} \tag{2.69}$$

The last expression is the required confidence interval for q. We now simply transform using $\rho_i = q/(q + 1)$:

$$\frac{F/F_U - 1}{n + F/F_U - 1} < \rho_i < \frac{F/F_L - 1}{n + F/F_L - 1} \tag{2.70}$$

Full details can be found in Searle *et al.* (1992: pp 65–6). Fleiss (1986) prefers to use a one-tailed interval of the form

$$\rho_i > \frac{F/F_L - 1}{n + F/F_L - 1} \tag{2.71}$$

where F_L now corresponds to $p = \alpha$, instead of $p = \alpha/2$.

Consider the map distances in Table 2.1. Our point estimate of ρ_i was 0.998. From the ANOVA results (Table 2.2) we find that $F_{9,40} = 2575.62$. Considering a one-sided 95% confidence interval (i.e. $\alpha = 0.05$), the corresponding F_L is 2.71. It follows that

$$\rho_i > \frac{(2575.62/2.71) - 1}{10 + (2575.62/2.71) - 1} = 0.990$$

Again, it is straightforward to use the jack-knife or bootstrapping (Section 2.7) to investigate the sampling properties of $\hat{\sigma}_\tau^2$ or of $\hat{\sigma}_e^2$, or the intra-class correlation, ρ_i, if one is not convinced of the normality assumptions used above. Typically, as replications are nested within subjects, we would be interested in, for example, deleting the whole data from each subject in calculation of the jack-knife estimates, or in sampling with replacement from the I subjects, in the case of bootstrapping. This is analogous to subsampling at the stage of the primary sampling unit in multistage survey sampling (see, for example Särndal *et al.*, 1992).

2.6.3 Variance components from the two-way ANOVA

Consider the following, where both subjects and raters are considered as random:

$$\text{Var}\left(\hat{\sigma}_\tau^2\right) = \text{Var}\left(\frac{SMS - EMS}{J}\right) = \text{Var}\left(\frac{SMS}{J}\right) + \text{Var}\left(\frac{SMS}{J}\right) \tag{2.72}$$

$$\text{Var}(\hat{\sigma}_\alpha^2) = \text{Var}\left(\frac{RMS - EMS}{I}\right) = \text{Var}\left(\frac{RMS}{I}\right) + \text{Var}\left(\frac{EMS}{I}\right) \qquad (2.73)$$

and

$$\text{Var}(\hat{\sigma}_e^2) = \text{Var}(EMS) \qquad (2.74)$$

The general form for these sampling variances is as follows:

$$\text{Var}\left(\sum_k w_k MS_k\right) = \sum_k \frac{2[w_k MS_k]^2}{v_k} \qquad (2.75)$$

where the MS_k (with d.f. v_k) is the kth mean square used to estimate a particular variance component, with corresponding weight w_k. An unbiased estimate of the variance component is obtained by replacing each of the v_k in this expression by $v_k + 2$ (see Searle et al., 1992; Brennan, 2001). In (2.72) and (2.73) both of the w_ks are $1/J$ and $1/I$, respectively. In (2.74) there is only a single mean square involved with weight equal to 1. If we regard raters as fixed, then the expressions for the sampling variances for each of the remaining variance components still apply. If we introduce subject by rater interaction then it will be treated as a random effect, irrespective of whether raters are regarded as fixed or random, and, again, (2.75) can be used to obtain the sampling variance of the estimated variance component. The sampling characteristics of variance components estimated via ML and REML will usually be obtained using second derivatives derived by iterative methods. It is straightforward to use the methods described in Section 2.6.1 to derive confidence limits for the error variance (σ_e^2), but not for the others. Tedious, but still approximate, methods can be found in Burdick and Graybill (1992) and Brennan (2001). Here we will discuss confidence interval estimation for two particular reliability coefficients.

For the two-way random effects model, without replication, Fleiss and Shrout (1978) use the Satterthwaite approximation (Satterthwaite, 1946) to provide an one-sided $100(1 - \alpha)\%$ confidence interval for a reliability coefficient defined by Equation (2.39) in the form of the following expression:

$$\rho_1 > \frac{N(SMS - F_L \times EMS)}{I \times SMS + F_L[J \times RMS + (IJ - I - J)EMS]} \qquad (2.76)$$

where F_L is the critical value of the F-distribution with v^* and $I - 1$ degrees of freedom, corresponding to the required value of α (typically 0.05). The term v^* is calculated as follows:

$$v^* = \frac{(I - 1)(J - 1)\left(J\hat{\rho}_i F + I[1 + (J - 1)\hat{\rho}_i] - J\hat{\rho}_i\right)^2}{(I - 1)J^2\hat{\rho}_i^2 F^2 + \left(I[1 + (J - 1)\hat{\rho}_i] - J\hat{\rho}_i\right)^2} \qquad (2.77)$$

where $F = SMS / EMS$. An alternative approximation has been provided by Arteaga et al. (1982). This is no simpler than the Shrout and Fleiss version, so we will not pursue it further here. See Burdick and Graybill (1992) and Brennan (2001) for further details.

What if we regard raters as fixed, and further that we are only interested in the reliability of measurements by a single rater (ie. the sampling distribution of the estimates of ρ_{1**} in Equation (2.48))? It is straightforward to show that

$$\hat{\rho}_{1**} = \frac{SMS - EMS}{SMS + (J-1)EMS} \tag{2.78}$$

and that a $100(1 - \alpha)\%$ one-sided confidence interval for ρ_{1**} is provided by

$$\rho_{1**} > \frac{F/F_L - 1}{J + F/F_L - 1} \tag{2.79}$$

where $F = SMS / EMS$ and F_L is the 100α percentage point of the F-distribution with $I - 1$ and $(I - 1)(J - 1)$ degrees of freedom (Fleiss, 1986; see also Feldt, 1965, Burdick and Graybill, 1992 and Brennan, 2001). We derived the reliability of Brian's guesses of string length in Section 2.5.2 ($\hat{\rho}_{1**} = 0.958$). The one-sided 95% confidence interval is given by

$$\rho_{1**} > \frac{\left(7.786/0.112\right)/1.91 - 1}{3 + \left(7.786/0.112\right)/1.91 - 1} = 0.922$$

If we abandon normality assumptions then we have to rely on methods such as the jack-knife or the bootstrap. But the application of these methods to cross-classified data involving two or more random effects is by no means simple and straightforward. We discuss these issues in the following two sections.

2.7   The jack-knife and bootstrap

Consider a simple random sample of measurements – the weights in Sections 1.1, for example. These were

$$490, 520, 500, 520, 470, 490, 490, 490, 500, 480.$$

Our aim in the analysis was to calculate the mean and the variance of these weights. The square root of the latter estimates the standard error of measurement (repeatability). In Section 2.6.1 we investigated the sampling distribution of this variance estimate under the assumption of normality of the measurement errors.

The jack-knife estimate of a standard error was introduced by Tukey in 1958 (see Mosteller and Tukey, 1977). The method is quite closely related to the bootstrap (Efron, 1979). Let $\hat{\theta}$ be the estimate of a particular parameter θ (s^2, for example), based on the whole data set of size n. Let $\hat{\theta}_{(-k)}$ be the estimate of θ when the kth observation is deleted from the sample. The kth pseudovalue, $\hat{\theta}^{(k)}$, is then calculated as

$$\hat{\theta}^{(k)} = n\hat{\theta} - (n-1)\hat{\theta}_{(-k)} \tag{2.80}$$

The jack-knife estimate of θ is then

$$\hat{\theta}^{(.)} = \frac{1}{n} \sum_{k=1}^{n} \hat{\theta}^{(k)}$$

(2.81)

and its sampling variance is given by

$$s^2(\hat{\theta}^{(.)}) = \frac{1}{n(n-1)} \sum_{k=1}^{n} \left(\hat{\theta}^{(k)} - \hat{\theta}^{(.)} \right)^2$$

(2.82)

The required standard error is then simply the square root of $s^2(\hat{\theta}^{(.)})$. The method is illustrated for the cook's ten weights in Table 2.11. The jack-knife standard error of the estimated standard error of measurement ($\hat{\sigma} = s$) is 3.66. Using normality assumptions (together with the delta technique – see Section 2.6.1) we obtain a corresponding estimate of 3.73.

Efron (1979) introduced a very general resampling procedure for investigating the sampling distributions of statistics based on an observed sample of observations of size n. This method is called the bootstrap. A bootstrap sample of size n is drawn at random, with replacement, from the original sample and the required statistic (a variance estimate, for example) obtained. The whole procedure is then repeated say, a thousand times, and the distribution of the bootstrap sampled statistics used to infer characteristics of the underlying but unknown sampling distribution. No distributional assumptions are made in any of the inferences – hence this form of bootstrapping is often referred to as the non-parametric bootstrap (a so-called parametric bootstrap will be introduced in Chapter 5). The standard deviation of the bootstrap estimates, for example, provides an estimate of the standard error of the statistic being investigated. The observed percentiles (2.5% and 97.5%, for example) provide simple to obtain confidence limits for the same statistic. Further details, and more subtle approaches to the use of the bootstrap, can be found in books by Efron and Tibshirani (1993), Shao & Tu (1995) and by Davison and Hinkley (1997). Recent practical guides are provided by Carpenter and Bithell (2000) and Wehrens *et al.* (2000).

Table 2.11 Jack-knife estimate of the standard error of the estimated standard error of measurement for the cook's weights.

Replicate	Weight	$\hat{\theta}_{(-k)}$	$\hat{\theta}^{(k)}$
1	490	16.667	8.107
2	520	13.944	32.614
3	500	16.667	8.107
4	520	13.944	32.614
5	470	13.944	32.614
6	490	16.677	8.017
7	490	16.677	8.017
8	490	16.677	8.017
9	500	16.677	8.017
10	480	15.811	15.811
		Mean:	16.194
		s.d.:	11.581
	Jack-knife estimate of se:		3.662

Returning to the cook's data, the results of taking 1000 bootstrap samples, and estimating the standard deviation of the weights for each of these bootstrap samples, led to an estimated standard error of the sample standard deviation of 3.36.

The application of the jack-knife or bootstrap to the one-way ANOVA model is straightforward if we think of our primary sampling units being the subjects (our pieces of string, for example). For the jack-knife we simply delete each subject (string) from the data set one at a time and repeat the estimation process to obtain the required pseudovalues. Similarly, our bootstrap estimates are based on sampling subjects (pieces of string) with replacement. This also applies in a similar way to two-way data when we regard the influence of raters, say, as fixed effects. Although we are ignoring variation at the level of the individual measurements, Brennan's Monte Carlo simulations indicate that this a reasonable approach (see Brennan, 2001; pp 201–7) – see below.

But what about two-way random effects models? Here the situation is considerably more complex. We, for example, have to consider random sampling of both subjects and raters. Cronbach et al. (1972) provide a jack-knife procedure for this model involving three different types of deletion of observations from the data. Considering the $I \times J$ matrix of observations, we first calculate the required variance component (or standard deviation), $\hat{\theta}$, using the whole data set, as before. We now delete all of the observations for each subject in turn and calculate the I values of $\hat{\theta}_{(-i)}$. Similarly we delete all of the observations for each rater in turn to obtain the J values of $\hat{\theta}_{(-j)}$. Finally, we delete each combination of the ith subject and jth rater in turn to obtain the IJ values of $\hat{\theta}_{(-ij)}$. The pseudovalue for the ith subject as measured by the jth rater is then given by

$$\hat{\theta}^{(ij)} = IJ\hat{\theta} - (I-1)J\hat{\theta}_{(-i)} - (J-1)I\hat{\theta}_{(-j)} + (I-1)(J-1)\hat{\theta}_{(-ij)} \qquad (2.83)$$

The jack-knife estimator of θ is simply the mean of the $\hat{\theta}^{(ij)}$, but the required standard error is a bit more complicated. We carry out a two-way analysis of variance on the IJ pseudovalues and estimate three variance components: $\hat{\sigma}^2_{rows}$, $\hat{\sigma}^2_{cols}$ and $\hat{\sigma}^2_{res}$. The required standard error is then given by

$$\sqrt{\frac{\hat{\sigma}^2_{rows}}{I} + \frac{\hat{\sigma}^2_{cols}}{J} + \frac{\hat{\sigma}^2_{res}}{IJ}} \qquad (2.84)$$

The reader is referred to Brennan et al. (1987) and Brennan (2001) for further details. We will not pursue the use of this method here as we will be primarily concerned with mixed effects models in the rest of this text.

For completeness, however, we will also briefly describe bootstrapping for two-way random effects models but, again, will not pursue the topic in any detail. Even if we were interested in two-way random effects modelling in the rest of the present text, it is clear from Brennan's simulation studies (Brennan, 2001; pp 201–7) that none of these bootstrap sampling procedures for two-way data perform very satisfactorily. Brennan (2001) bases his discussion on his own earlier work (Brennan et al., 1987) and an unpublished PhD thesis by Wiley (Wiley, 2000). For a two-way design involving I subjects and J raters, he considers four sampling procedures:

A. Draw a sample of I subjects with replacement and then draw an independent random sample of J raters. The bootstrap sample consists of the measurements for the sampled subjects by the sampled raters.

B. Regard the raters as fixed and simply draw a sample of I subjects. Then the measurements on the selected subjects comprise the required bootstrap sample.

C. Regard the subjects as fixed and simply draw a sample of J subjects. Then the measurements by the selected raters comprise the bootstrap sample.

D. From the original IJ measurements, estimate the subject effect (as a deviation from the overall mean), the rater effect (as a deviation from the overall mean) and the residual effect (the deviation of the ijth measurement from the sum of the overall mean, the ith subject effect and the jth rater effect). The bootstrap sampling procedure then involves (a) sampling with replacement from the I subject effects, (b) independently sampling with replacement from the J rater effects, and (c) independently sampling with replacement from the IJ residuals. The ijth bootstrapped observation is then the sum of the three independently sampled components.

Using normally-distributed simulated data with the number of subjects (I) equal to 100 and the number of raters (J) equal to 20, Brennan (2001) found that the jack-knife procedure worked satisfactorily but that the results from bootstrapping were rather mixed. Method A seemed to work reasonably well for σ_r^2 (i.e. the variance of the rater effects). Method B was OK for both σ_τ^2 (i.e. the variance of the subject effects) and σ_e^2 (the 'error' variance). Method C was OK for σ_r^2 and even better for σ_e^2. Finally, the most complex procedure, Method D, worked reasonably well for σ_r^2 and for σ_e^2. The correct interpretation of these findings is not obvious. Brennan, however, concludes that since the two-way random effects design is symmetric,

> at least some of the asymmetry in these results is attributable to the five-to-one ratio of sample sizes. Apparently, the applicability of the bootstrap depends to some extent on both the bootstrap sampling procedure employed and the pattern of sample sizes.

My own conclusion is that it is probably safe to use the jack-knife, but that bootstrap sampling does not appear to be sufficiently developed and understood under these circumstances, and that great care should be taken in its routine use. In fact, there probably should not be any routine use of the bootstrap for these two-way models at present! One obvious point, however, is that if the investigator is seriously interested in the estimation of random rater effects, irrespective of the method of data analysis to be used, then the design of the study must involve more than a handful of raters (and a lot more than the two involved in the great majority of published reliability studies!).

Various options for increasing the number of raters in a two-way generalizability study of this sort, without making the study entirely impractical or prohibitively expensive, are discussed in Dunn (1992; Section 5.3). Options include balanced incomplete blocks designs (Fleiss, 1981a, 1986), generalized staggered designs (Bush and Anderson, 1963), disjoint rectangle designs (Gaylor, 1960) and various forms of nesting (see, for example, Goldsmith and Gaylor, 1970). In my 1992 review I suggested trying various hybrid designs. The only example of real data that I am aware of can be found in Macnab *et al.* (1994). These are shown in Table 2.12. The study involved ratings of live and videotaped interviews for 16 children using the Vancouver sedative recovery scale. Raters comprised of two experts and 16 novices. The two experts independently assessed all the live interviews for all 16 children. The 16 novices each independently assessed six videotapes selected in accordance with a BIBD design. Each of the tapes was assessed by six raters, and each pair of raters scored two children in common. The data will be analysed in the following section.

Table 2.12 Rating data (novice and expert) for 16 cases in an incomplete blocks design.

Raters	|	Cases 1	2	3	4	5	6	7	8	9	10	11	12	13	14	15	16
1	|				6				0	18		12		14			19
2	|				5		4			13	13		0			16	
3	|			20		13	1		0	15					15		
4	|	9						22		19	18				14	14	
5	|	11	0	21						19		10				19	
6	|		1			13		22		16			0				22
7	|	12	0		0		1	22	0								
8	|		1	19	5	14						15		14			
9	|				9	16		15				10			12	19	
10	|	10		17	7								0		13		20
11	|		0					0					0	13	16	22	
12	|			20				20	0		12	10	0				
13	|	6				15			0		17					19	22
14	|	8				16	1					11	0	15			
15	|			21			1	18						16		21	22
16	|		0				1					18	15		10		21
17	|	10	0	21	8	18	4	16	0	18	16	10	0	16	12	19	20
18	|	10	0	21	5	19	7	18	0	16	15	12	0	14	20	20	21

Raters 1–16 are the novices; 17 and 18 are the experts.
Source: Macnab *et al.*, 1994.

2.8 Coping with missing data

Once we have missing observations, whether by design or by accident, the attractions of the ANOVA methods (i.e. equating observed and expected mean squares) soon disappear. Fleiss (1986) provides an example of a fairly easy-to-use approximation in the case of the one-way model in which there are a variable number of replicated measurements available for each of the sampled subjects, and he has also provided details of how to use these ANOVA-based methods for data obtained using a BIBD (Fleiss, 1981a, 1986). In this situation we will usually assume normality and use either ML or REML to estimate parameter values and their standard errors. As long as we can assume that the missing data mechanism is non-ignorable (i.e. either missing completely at random or missing at random in the sense of Little and Rubin, 2002) then ML or REML methods can be used without too much worry. Other estimation methods for unbalanced data include Henderson's adaptations of ANOVA methods (Henderson, 1953), minimum norm unbiased estimation (MINIQUE – LaMotte, 1973; Rao, 1972) and minimum variance quadratic unbiased estimation (MIVIQUE – Hartley *et al.*, 1978). Interested readers are referred to Searle *et al.* (1992) for further details. If the missing data are not too extensive it might be of interest to consider multiple imputation (Little and Rubin, 2002) in conjunction with the standard ANOVA methods.

Let's have a look at the data in Table 2.12 using REML. Just scanning the data values shows that the experts do not agree perfectly, but their assessments for each child do appear to be closer (less variable) than those of the novices. The variation between children appears to be much larger than that due to rater or measurement errors. If we analyse the results for the two experts on their own we obtain the following estimates: $\hat{\sigma}_\tau^2 = 54.89$ (se 20.29); $\hat{\sigma}_r^2 = 0$; $\hat{\sigma}_e^2 = 1.31$ (se 0.46). If we look at the novices on their own we obtain: $\hat{\sigma}_\tau^2 = 58.25$

(se 21.56); $\hat{\sigma}_r^2 = 0$; $\hat{\sigma}_e^2 = 4.84$ (se 0.77). Finally, if we look at the whole data set in a single analysis we obtain: $\hat{\sigma}_\tau^2 = 57.09$ (se 21.04); $\hat{\sigma}_r^2 = 0.024$ (se 0.227); $\hat{\sigma}_e^2 = 4.25$ (se 0.61). There is a suspicion that the novices are less precise than the two experts ($\hat{\sigma}_e^2 = 4.84$ as opposed to 1.31, respectively), but little evidence of any biases. It's encouraging to see that the analysis of the whole data set brings the estimate of the rater variance component away from the boundary (i.e. zero). The reliability estimate based on the whole data set is 0.930. As yet, we have not covered models that might enable us to test for differences in precision between the novices and experts (or even whether each of the novices is equally precise). This is covered in Chapter 4.

3

Method comparison 1: paired observations

3.1 Introduction: what is the purpose of a method comparison study?

In a measurement comparison study, each of a sample of n subjects or specimens provides a measurement using each of K different instruments or techniques. Frequently there is no replication of the measurements by each of the instruments and so, subject to there being no missing data, we are concerned with the analysis of nK data points. In the more informative designs, however, the study will involve repeated measurements using some if not all of the K instruments. In the present chapter we primarily consider the most commonly used design in which we wish to compare the performance of only two methods or instruments and in which there is no replication (i.e. we have, at most, $2n$ observations). In many ways it is unfortunate that this design is so common because the resulting data are often difficult to interpret with confidence. The lack of replication is not necessarily a design fault, however. In an early statistical paper on this subject, Grubbs was interested in the burning times of fuses for explosives (Grubbs, 1948) and later he discussed the measurement of the velocity of artillery shells (Grubbs, 1973). In both cases we only get the opportunity to observe the process once! Grubbs' fuse burning data are shown in Table 3.1. They do not actually conform to the design considered in the present chapter since in both cases $K = 3$. This situation will be covered in detail in the following chapter.

The motivation of method comparison studies is often the desire to replace a difficult or expensive measurement procedure with an easier or cheaper one. Lewis *et al.* (1991) distinguished at least three different classes of measurement comparison study:

(a) *Calibration problems* arise when the comparison is between an approximate method and a known accurate and precise method. The object here is to establish a relationship between the two methods which maps a measurement made using the approximate method to a known probability distribution of true measurements. This can often be achieved using standard regression techniques.

(b) *Comparison problems* arise when the comparison is between two methods measuring in the same units, neither of which is thought to be accurate or precise. In this case, it is important to know to what extent the results of the methods agree and to understand the nature of any difference between them. …

Table 3.1 Fuse burning times.

Observer A	Observer B	Observer C	Observer A	Observer B	Observer C
10.10	10.07	10.07	9.74	9.73	9.74
9.98	9.90	9.90	10.32	10.32	10.34
9.89	9.85	9.86	9.86	9.86	9.86
9.79	9.71	9.70	10.01	lost	10.03
9.67	9.65	9.65	9.65	9.64	9.65
9.89	9.83	9.83	9.50	9.49	9.50
9.82	9.75	9.79	9.56	9.56	9.55
9.59	9.56	9.59	9.54	9.53	9.54
9.76	9.68	9.72	9.89	9.89	9.88
9.93	9.89	9.92	9.53	9.52	9.51
9.62	9.61	9.64	9.52	9.52	9.53
10.24	10.23	10.24	9.44	9.43	9.45
9.84	9.83	9.86	9.67	9.67	9.67
9.62	9.58	9.63	9.77	9.76	9.78
9.60	9.60	9.65	9.86	9.84	9.86

Source: Grubbs (1948).

(c) *Conversion problems* arise when the comparison is between two approximate methods of measurement each of which measures the quantity in different units. This situation can arise when the methods in question proceed by measuring different proxies for the underlying quantity of interest.

St Laurent (1998) has added a fourth type:

(d) gold-standard comparison problems, in which an approximate method of measurement is compared to a gold standard in order to assess the degree to which the approximate method agrees with the gold standard.

By using the phrase 'gold standard' St Laurent is referring to the method known to be accurate and precise as described in Lewis *et al.*'s calibration study described above. Although this classification is useful, particularly when we begin to look at the controversies over the choice of method of analysis, readers should not assume that there are sharp and clear-cut boundaries between the four types. No measurement method is infallible (error-free) and so it is of necessity a subjective decision when we come to decide that a particular method or instrument can be treated as if it were a gold standard. Although two or more methods might be nominally scaled using the same units (in the comparison studies as defined by Lewis *et al.*) there is always the possibility of biases. Mathematically, bias cannot be distinguished from scale differences. Adjustment for bias is equivalent to scale conversion. So, really, the distinction between the comparison problem (b) and that of conversion (c), and also between (a) and (c), is just a matter of degree.

The choice of method of statistical analysis of method comparison data seems to be the subject of endless controversy, particularly in areas of application such as clinical biochemistry and diagnostic medicine. Much of this controversy appears to arise from the frequently rather vague or poorly stated aims of the investigators (or even their complete absence!). So, before going into alternative methods of statistical analysis in any detail it is vital that we get some idea of what the aims of a method comparison study actually are. I will start with a few quotations from the relevant literature.

The question being answered is not always clear, but is usually expressed as an attempt to quantify the agreement between the two methods.

(Bland & Altman, 1995)

Some lack of agreement between different methods of measurement is inevitable. What matters is the amount by which methods disagree. We want to know by how much the new method is likely to differ from the old, so that if it is not enough to cause problems in the mathematical interpretation we can replace the old method by the new, or even use the two interchangeably.

(Bland & Altman, 1999)

It often happens that the same physical and chemical property can be measured in different ways. For example, one can determine sodium in serum by flame atomic emission spectroscopy or by isotope dilution mass spectroscopy. The question then arises as to which method is *better.*

(Mandel 1991)

In areas of inter-laboratory quality control, method comparisons, assay validation and individual bio-equivalence, etc., the agreement between observations and target (reference) values is of interest.

(Lin, 2000)

The purpose of comparing two methods of measurement of a continuous biological variable is to uncover systematic differences, not to point to similarities.

(Ludbrook, 1997)

In the pharmaceutical industry, measurement methods that measure the quantity of products are regulated. The FDA (U.S. Food and Drug Administration) requires that the manufacturer show equivalency prior to approving the new or alternative method in quality control.

(Tan & Inglewicz, 1999)

There appears to be a common core to this controversy: are we concerned in the demonstration of equivalence of two or more measurement techniques (for example, by adequate agreement), or are we more interested in finding differences (the best)? Are the two views mutually exclusive? If we are primarily interested in the problem of equivalence, how do we assess this equivalence? Is equivalence the same as agreement? If, however, we are asking which of a set of alternative measurement methods is the better, how do we define 'better'? Is the best method the one that is the most precise? What about its cost and relative ease of use? On the assumption that two instruments or methods are measuring the same thing (the method under test has perfect *validity*, with respect to the standard), does the question of equivalence boil down to asking whether their precisions are near enough the same? Or, should we be asking whether the precision of one instrument is demonstrably better than that of the other?

A particularly influential contribution to this debate is the work of Bland and Altman (Altman & Bland, 1983; Bland & Altman, 1986). These two authors have quite forcibly argued that investigators should be investigating agreement between measurement methods with the view to asking 'Do the two methods of measurement agree sufficiently closely?' (Altman & Bland, 1983). In trying to prevent the widespread misuse of statistical methodology these authors have gone a bit too far. Although the simple methods that have been advocated by Bland and Altman (see the following two sections for details) are a big improvement over the naïve use of, say, correlation coefficients or simple linear regression, they are not always appropriate or particularly informative and, given data from a sufficiently sophisticated design, are totally inadequate for the exploration of differences in the scaling and precision of the competing measurement methods. It seems quite reasonable to ask, for example, whether one method is better (more precise, say) than another. If it were not for improvements in technique we would still be measuring length by referring to the sizes of our hands, arms or feet! The clinicians' so-called gold standard (measurement of blood pressure using a mercury sphygmomanometer, for example) often leaves considerable room for improvements.

Bland and Altman concentrate on the estimation of limits of agreement (see Section 3.3) in their assessment of whether two methods might be equivalent, for example, in the sense of being used interchangeably. But this notion of equivalence is too restrictive. I would be quite happy to exchange a high quality thermometer calibrated in °Fahrenheit for one of comparable quality calibrated in °Celsius. These two instruments could easily and should be regarded as equivalent, but they will never agree with each other in the restrictive sense used by Bland and Altman. Agreement implies equivalence, but not *vice versa*. If, however, two methods are equivalent then appropriate re-scaling (conversion) will bring the two sets of measurements into agreement. In this context a simple correlation coefficient is likely to be both more informative and valid as an index of equivalence than a function of the standard deviation of the differences between the two sets of measurements. And if we are interested in re-scaling (conversion) then it is also inevitable that we need to use some sort of regression analysis (although not the simple ordinary least-squares regressions of the naïve practitioners).

Like Bland and Altman here we are often concentrating on the demonstration of equivalence. Unlike Bland and Altman, however, we do not restrict ourselves to simply dealing with agreement. We are concerned with the (usually) joint estimation re-scaling factors (regression coefficients) and precision (variance estimates). For this purpose our main tool is the use of measurement error modelling (Fuller, 1987; Cheng & Van Ness, 1999), supplemented by the use of preliminary graphical techniques and simple summary statistics. In the pages of the present text the following view is rejected almost in its entirety:

> Our approach is based on the philosophy that the key to method comparison studies is to quantify disagreements between individual measurements. It follows that we do not see a place for methods of analysis based on hypothesis testing. ... Widely used statistical approaches which we think are misleading include correlation, regression, and the comparison of means. Other methods which we think inappropriate are structural equations and intra-class correlation.
>
> (Bland & Altman 1999)

We will discuss and illustrate the use of all the methods criticised above! It is not the methods *per se* that should be criticised in this way, but the way they might have been used by ill-informed and naïve investigators.

If we are considering a simple comparison problem, say, of measurements X and Y, in which the scales of measurement of the two methods are the same (and there are no systematic biases – τ-equivalent in the language of the psychometrician – see Section 1.3) then the two devices are equivalent (parallel psychometric tests, for example) if the variance of the measurement errors in Y (say, σ_ε^2) is equal to that of the standard, X (say, σ_δ^2). That is, they are equivalent if the precision ratio (the ratio of the two variances), $\lambda = \sigma_\varepsilon^2/\sigma_\delta^2$, equals unity. If, however, our joint measurement model for X and Y is

$$X_i = \tau_i + \delta_i$$
$$Y_i = \alpha + \beta\tau_i + \varepsilon_i \tag{3.1}$$

where the δ_i and ε_i are uncorrelated measurement errors, then the two methods are equivalent if $\psi = \sigma_\varepsilon^2/\beta^2\sigma_\delta^2 = \lambda/\beta^2 = 1$. Note that if we were to assume that the standard method is infallible (i.e. $\sigma_\delta^2 = 0$) then ψ would be undefined. If there are random subject- or item-specific biases (matrix effects) as described in Section 1.4, we have to modify these two indices of equivalence accordingly (that is, σ_ε^2 is replaced by $\sigma_{sy}^2 + \sigma_\varepsilon^2$, where σ_{sy}^2 is the variance of the item-specific biases for Y, for example, and, similarly $\sigma_{\varepsilon x}^2$ is replaced by $\sigma_{xy}^2 + \sigma_\delta^2$).

Now, we are not usually concerned with whether ψ (or λ) is exactly 1. If the new method were suitable as a replacement for the standard we would like to show that ψ (or λ) is below a certain threshold. This threshold might be set at 1.3, for example. If ψ (or λ) is above 1.3 then the precision of the new method is unsatisfactory. If it is below 1.3 then it is either equivalent to the standard or better. If, however, we wish to establish that the new method is better (more precise) than the standard then we would use a threshold less than 1. We might want it to be less than 0.7, for example. Equivalence might be taken to be indicated by a value of ψ (or λ) in the range (0.7, 1.3). The choice of these thresholds depends on the substantive knowledge of the investigator and the context of the use of the measurements.

An investigator interested in establishing that his new method is better than the standard is likely to be concerned with a 95% confidence interval for ψ (or λ) which does not include ψ (or λ) $= 1$ and is in the range of ψ (or λ) which is lower than 1. It might be either a one- or two-sided (the upper 95% one-sided confidence limit is 0.75, for example, or the 95% two-sided confidence interval is (0.55, 0.70)). In terms of statistical significance tests, the null hypothesis is H_0: ψ (or λ) $= 1$. The alternative is H_A: ψ (or λ) < 1.

An investigator interested in a formal significance test of equivalence has a null hypothesis that the methods are *not* equivalent. That is, H_0: ψ (or λ) < 0.7 or ψ (or λ) > 1.3. His alternative hypothesis (H_A) is that ψ (or λ) is within the range (0.7, 1.3). In terms of a two-sided confidence interval he requires an interval that lies wholly within the range (0.7, 1.3). It is important to recognise that failure to demonstrate that ψ (or λ) is different from 1 is *not* the same as demonstrating satisfactory equivalence. If we are simply looking to fail to demonstrate that ψ (or λ) is different from 1 then all we need is a very small sample! In order to demonstrate equivalence we would need a large sample, and frequently a very large one.

We illustrate the problem by reference to the motivation for a recent paper of Tan and Inglewicz (1999):

> A recent dataset from Merck (Merck & Co., Inc.) provides an example. A high performance liquid chromatography (HPLC) method was developed several years ago and filed with the FDA for testing the quality of Famitidine tablets (a histamine H_2-receptor antagonist for inhibition of gastric secretion). Later, it was found that this method was not selective for some possible degradates. A new, modified HPLC method was then developed (new column and different injection volume). The FDA requested that Merck show equivalency, in terms of quantification of the active incredient Famatidine, between the originally filed method and the new method prior to use of the new method in manufacture quality-control testing.

The aim of Tan and Inglewicz's subsequent investigation was to estimate the parameters of a model such as that in Equation (3.1) – that is, to estimate α, β, σ_δ^2 and σ_ε^2 – and to use these estimates to establish what they define to be individual equivalence ($\lambda = 1$, $\beta = 1$, $\alpha = 0$), average equivalence ($\beta = 1$, $\alpha = 0$) or sensitivity equivalence ($\psi = 1$, $\alpha = 0$). Individual equivalence is equivalent to parallellism in psychometrics, whereas average equivalence is the same as the psychometrician's τ-equivalence. The relative sensitivity of

Y with respect to X (as in the phrase 'sensitivity equivalence') is defined, in the context of a model such as (2.1), by Mandel (1991) to be $\beta/\sqrt{\lambda}$. Our measure of equivalence (ψ) is the square of Mandel's sensitivity (N.B. do not confuse this use of the word with other uses of the same word in this text – see Section 1.7 and Chapter 5). Similar ideas of equivalence have been discussed by Hawkins (2002). In the main part of this chapter we will be explaining the use of measurement error modelling in the point and interval estimation of α, β, λ, and ψ, together with the use of associated significance tests. We start with two sections covering graphs and other simple descriptive techniques.

Before leaving this section, however, we should add one further motivation for the conduct of method comparison studies. We are interested in the pattern of measurement errors (both systematic and random) and we use formal statistical models as a means of evaluating these patterns and getting insight into the measurement process. Statistical modelling does not always gives us elegant and convincing answers (in fact, this appears to be the exception rather than the rule in the use of measurement error modelling!) but the process of trying and rejecting models will lead us to think more deeply about what might be going on (or not) during the measurement process.

3.2 Preliminary numerical and graphical summaries

Let's assume that we have carried out a simple method comparison study in which we have carried out measurements on each of n specimens or subjects using $K = 2$ methods. We have no replication within methods. We are (or should be) interested in the marginal distributions of the measurements within each of the two methods separately, and also in their pattern of covariation. Table 3.2 provides data on the weights (in grammes) of 15 packets of potatoes (items), each measured using one of two kitchen scales (Scale A and Scale B). Also given in this table are the differences (A − B) and means ((A + B)/2). Table 3.3 displays several simple summary statistics for these data (variances, covariances, etc. being estimated using the divisor $n - 1$). Figure 3.1 shows a simple two-way scatter diagram of A plotted against B.

Table 3.2 Comparison of two kitchen scales (weights in grammes).

Item	Scale A	Scale B	A − B	(A + B)/2
1	135	165	−30	150
2	940	910	30	925
3	1075	1060	15	1067.5
4	925	925	0	925
5	2330	2290	40	2310
6	2870	2850	20	2860
7	1490	1425	65	1457.5
8	2110	2050	60	2080
9	650	630	20	640
10	1380	1370	10	1375
11	970	1000	−30	985
12	1000	1000	0	1000
13	1640	1575	65	1607.5
14	345	345	0	345
15	310	320	−10	315

What can we learn from these simple summaries? Both the means and standard deviations of Scales A and B are very similar, and the distribution of both sets of measurements seems to be reasonably symmetrical. They are also highly correlated, as indicated from the scatter diagram and the estimated correlation of 0.9995. From the scatter diagram, there is little evidence that the relationship between the two scales is other than linear, with a slope approximately equal to 1.0. There do not appear to be any outlying or discrepant observations.

Table 3.3 Simple summaries of the kitchen scales data in Table 3.2.

Univariate summary statistics

	Scale A	Scale B	A − B	(A + B)/2
Obs.	15	15	15	15
Mean	1211.333	1194.333	17.000	1202.833
Std. dev.	775.227	756.573	30.752	765.803
Min.	135	165	−30	65
Max.	2870	2850	65	2860

Covariance matrix

	A	B
A	600977	
B	586217	572403

Corr (A, B)   = 0.9995
Corr (A − B, (A + B)/2) = 0.6066

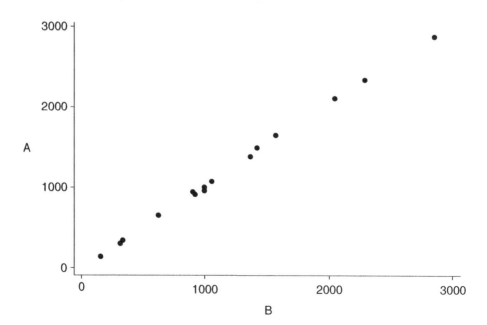

Figure 3.1   Weights of potatoes: Scale A versus Scale B.

Finally, there does not appear to be a large (constant) systematic difference between the two sets of measurements (the mean difference is only 17 g). But is there any evidence that the difference between the two scales might depend on the size of the packet being weighed (estimated by the mean of A and B)? The correlation between $A - B$ and $(A + B)/2$ is 0.6066. A plot of A against $(A + B)/2$ – the Bland–Altman plot (Altman & Bland, 1983; Bland & Altman, 1986) – is shown in Figure 3.2. It appears that, on average, the difference $(A - B)$ is increasing with increasing mean, but we should be careful about jumping to conclusions with so few data. But what might be the explanation for such a correlation? One possibility is that the scale of measurement of A and B, although similar, are not identical (i.e. $\beta \neq 1$). The other is that, although $\beta = 1$, the precisions of the two scales are not the same (see Section 3.7 to see how this phenomenon induces the observed correlation). And these two possible explanations, of course, are not mutually exclusive. A final plot is shown in Figure 3.3. For each item we have calculated the ratio A/B and then plotted this ratio against the mean, $(A + B)/2$ (see Eksborg, 1981; Bland & Altman, 1999). Again, it looks as if the ratio (a rough estimator of β if the A vs. B plot passes through the origin) is approximately 1.0.

Table 3.4 compares the results of using high performance liquid chromatography (HPLC) with a monoclonal antibody-mediated flourescence polarization immunoassay (FPIA) in the measurement of the concentrations of cyclosporine (CsA) in 20 selected kidney samples (i.e. the first 20 samples from Table 2 of Napoli *et al.*). Table 3.5, and Figures 3.4 to 3.6, illustrate the characteristics of these data. The conclusions are essentially the same as those reached in the interpretation of the weights of packets of potatoes, but with one difference. There appears to be an outlier (item 15). This can most easily be seen from the Bland–Altman plot (Figure 3.5) and the ratio vs. mean plot (Figure 3.6). We will not exclude

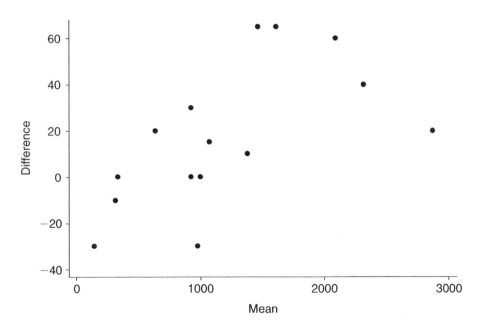

Figure 3.2   Weights of potatoes: Difference $(A - B)$ versus the mean, $(A + B)/2$.

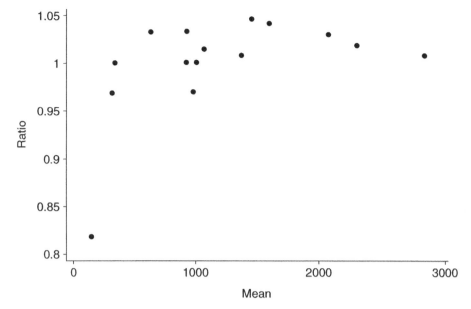

Figure 3.3   Weights of potatoes: Ratio (A/B) versus the mean, (A + B)/2.

Table 3.4 Kidney concentrations of CsA (ng/200 mg tissue wet weight).

Item	HPLC	FPIA	Item	HPLC	FPIA
1	104	112	11	352	326
2	854	804	12	272	206
3	682	559	13	250	132
4	194	234	14	242	278
5	248	299	15	532	219
6	308	266	16	216	143
7	216	287	17	912	981
8	70	113	18	320	218
9	546	664	19	234	152
10	364	259	20	850	869

Source: Napoli *et al.* (1996).

this observation in any subsequent analysis but note that Napoli *et al.* (1996) excluded items when the ratio of the HPLC to FPIA measurements was greater than 2. Although there appears to be little evidence of heteroscedasticity of error variances in these data (increase in the variability of the HPLC − FPIA differences with increasing mean), Napoli *et al.* state that there are *a priori* reasons to expect that the variance of the errors would increase in direct proportion to the mean.

For the final example in this section, let us have a look at the data used to illustrate problems of scale conversion by Lewis *et al.* (1991). Table 3.6 gives the raw data on the indirect measures of fuel consumption in 120 consecutive family car trips. Summary statistics are provided in Table 3.7. Note that the data are highly positively skewed. Table 3.7 also provides comparable summary statistics for the logarithms of the measurements. Graphical summaries for the raw data can be found in Figures 3.7, 3.8 and 3.9, and those for the

Table 3.5 Summary statistics for the kidney CsA concentration data.

Variable	Obs	Mean	Std. Dev.	Min	Max
HPLC	20	388.3	254.0122	70	912
FPIA	20	356.05	266.9533	112	981
Diff	20	32.25	97.4668	−118	313
Mean	20	372.175	255.9652	91.5	946.5

After excluding item 15

Variable	Obs	Mean	Std. Dev.	Min	Max
HPLC	19	380.74	258.6488	70	912
FPIA	19	363.26	272.2586	112	981
Diff	19	17.47	73.6081	−118	123
Mean	19	372.00	262.978	91.5	946.5

Corr (HPLC, FPIA) = 0.9312
Corr (HLPC, FPIA) = 0.9628 after excluding item 15

Covariance matrix ($n = 20$)

	HPLC	FPIA
HPLC	64522.2	
FPIA	63143.2	71264.1

Covariance matrix ($n = 19$;
i.e. excluding item 15)

	HPLC	FPIA
HPLC	66899.2	
FPIA	67802.9	74124.8

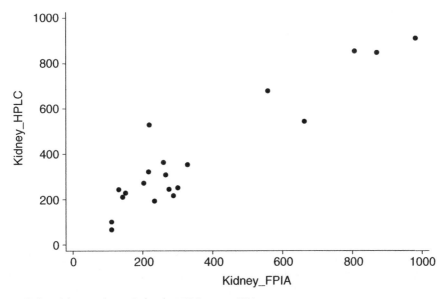

Figure 3.4   Kidney cyclosporin levels: HPLC versus FPIA.

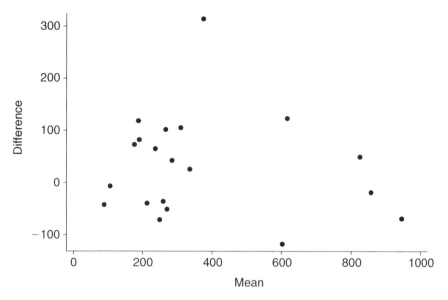

Figure 3.5 Kidney cyclosporin levels: Difference (FPIA − HPLC) versus the mean, (FPIA + HPLC)/2.

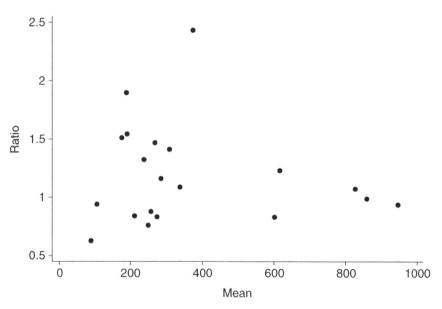

Figure 3.6 Kidney cyclosporin levels: Ratio (FPIA/HPLC) versus the mean, (FPIA + HPLC)/2.

log-transformed data are in Figures 3.10 and 3.11. Figure 3.11, for example, is a Bland–Altman plot based on the logarithms of the measurements, rather than the raw data themselves. Interpretation is not simple! Lewis *et al.* chose to base their analysis on transformed measurements.

Table 3.6 Fuel consumption measurements for 120 consecutive car trips.

Journey	Turbine meter (cm^3)	Displacement meter (ml)	Journey	Turbine meter (cm^3)	Displacement meter (ml)
1	682	49	61	62204	5102
2	4803	356	62	605	46
3	736	50	63	887	65
4	3574	269	64	6332	488
5	3905	308	65	2925	216
6	6959	543	66	1469	113
7	5062	401	67	4019	301
8	729	44	68	103264	11276
9	3760	251	69	36114	3046
10	1862	143	70	69902	6005
11	4245	300	71	697	47
12	2654	181	72	1299	106
13	867	55	73	3781	291
14	5724	385	74	4138	316
15	1151	79	75	5624	453
16	1085	78	76	919	75
17	65838	5032	77	1714	134
18	91491	7680	78	2309	185
19	498	33	79	4358	333
20	1768	136	80	5895	476
21	9322	760	81	23039	1779
22	49474	4115	82	3944	322
23	36674	3073	83	4435	354
24	63624	4983	84	5549	423
25	666	44	85	969	81
26	3883	322	86	5143	407
27	5197	299	87	2844	198
28	2338	188	88	9244	669
29	6055	434	89	4853	359
30	2120	161	90	4348	336
31	6294	420	91	820	53
32	7467	516	92	5021	392
33	1162	93	93	3358	255
34	839	60	94	1784	140
35	4610	296	95	5839	472
36	4291	322	96	5015	355
37	119843	10994	97	5241	417
38	4464	320	98	1508	109
39	33147	2750	99	3250	255
40	69353	5951	100	3146	237
41	4485	312	101	4001	297
42	5131	358	102	6943	521
43	7022	357	103	1714	128
44	6107	424	104	3506	248
45	4600	352	105	5216	412
46	1574	109	106	1748	134
47	5017	387	107	1253	88
48	5730	453	108	5328	428
49	5209	406	109	4404	327
50	1360	93	110	2990	230
51	3986	343	111	4432	328
52	4584	344	112	1650	124
53	62371	5225	113	4603	377
54	24475	2167	114	930	78
55	57544	4845	115	19176	1504
56	496	29	116	21843	1747
57	925	76	117	4310	323
58	2488	152	118	1609	132
59	48671	3984	119	914	67
60	37991	3281	120	3693	272

Source: Lewis *et al.* (1991).

Table 3.7 Summary statistics for fuel consumption measurements (Table 3.6).

(a) Raw data

Variable	Mean	Std. dev.	Min	Max
Turbine	12050.42	22220.44	496	119843
Displace	1005.192	2000.913	29	11276
Correlation	0.993			

(b) Logarithms of measurements

Variable	Mean	Std. dev.	Min	Max
Log(Turbine)	8.400	1.296	6.207	11.694
Log(Displace)	5.816	1.349	3.367	9.330
Correlation	0.998			

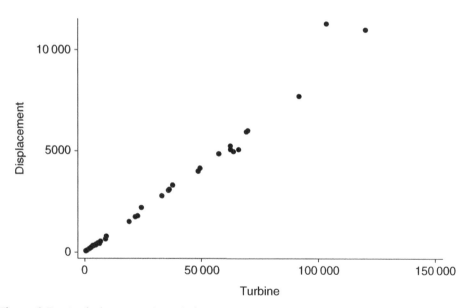

Figure 3.7   Car fuel consumption: Displacement meter versus Turbine meter.

3.3 Indices of concordance or agreement

For a pair of measurements (X_i, Y_i) on an individual specimen or subject the obvious measure of their disagreement is simply the difference between them $(d_i = X_i - Y_i)$. X_i and Y_i can be used interchangeably if the difference between X_i and Y_i is sufficiently small (what is meant by 'sufficiently small' is determined by substantive criteria rather than purely statistical ones). Bland and Altman have proposed the use of a *reference range* for these differences based on the interval that contains, say, 95% of the differences on repeated measurement (either on the same subject or on different ones). This interval they call the

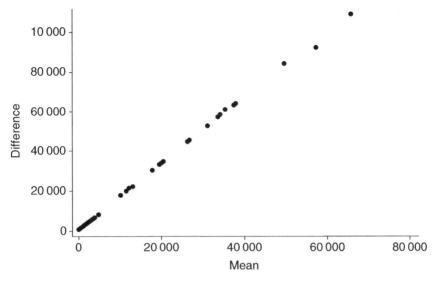

Figure 3.8 Car fuel consumption: Difference (Turbine − Displacement) versus the mean, (Turbine + Displacement)/2.

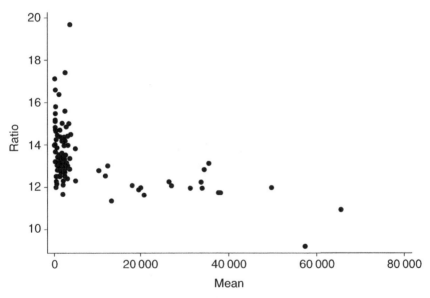

Figure 3.9 Car fuel consumption: Ratio (Turbine/Displacement) versus the mean, (Turbine + Displacement)/2.

95% *limits of agreement* (see, for example, Bland & Altman, 1999). It is estimated from the data provided by the method comparison study. On the assumption of the normality of the differences, it is simply the sample mean of the differences plus or minus 1.96 times the estimated standard deviation of the differences (i.e. $\bar{d} \pm 1.96S_d$). A distribution-free (and usually more robust) version is simply based on the 2.5th and 97.5th percentiles of the

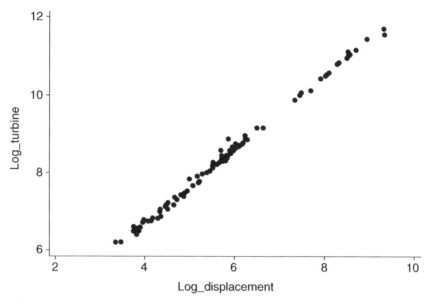

Figure 3.10 Car fuel consumption: Logarithm of turbine meter versus logarithm of dispacement meter.

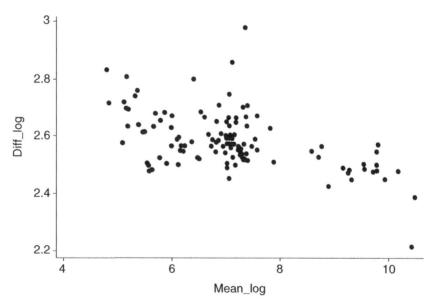

Figure 3.11 Car fuel consumption: Difference, log(turbine) − log(displacement) versus the mean, [log(turbine) + log(displacement)]/2.

cumulative distribution function for the differences and, again, this can be estimated from the data (Bland & Altman, 1999). Returning to the normally distributed differences:

$$\mathrm{Var}(\bar{d} \pm 1.96\,S_d) = \mathrm{Var}(\bar{d}) + 1.96^2\,\mathrm{Var}(S_d) \qquad (3.2)$$

where

$$\text{Var}(S_d) \simeq \frac{S_d^2}{2(n-1)} \tag{3.3}$$

For further details, see Bland & Altman (1999). For reasonably large n:

$$\text{Var}(\bar{d} \pm 1.96\,S_d) \simeq 1.71^2 \frac{S_d^2}{n} \tag{3.4}$$

The standard errors of $\bar{d} - 1.96\,S_d$ and $\bar{d} + 1.96\,S_d$ are therefore approximately $1.71\,S_d/\sqrt{n}$. A 95% confidence interval for the lower and upper limits of agreement can then be calculated by finding the approximate value of the t distribution with $n - 1$ degrees of freedom. The confidence limits will be t standard errors either side of the observed value.

Another natural contender for the degree of discordance between two measurements (X_i, Y_i) is the square of the difference between them (i.e. $d_i^2 = (X_i - Y_i)^2$). For the sample as a whole:

$$\text{Observed discordance} = \sum_i d_i^2 \tag{3.5}$$

The expected value of d_i^2 is given by

$$\begin{aligned}
E(X_i - Y_i)^2 &= (\mu_X - \mu_Y)^2 + (\sigma_X^2 + \sigma_Y^2 - 2\sigma_{XY}) \\
&= (\mu_X - \mu_Y)^2 + (\sigma_X^2 + \sigma_Y^2 - 2\rho\sigma_X\sigma_Y)
\end{aligned} \tag{3.6}$$

where ρ is the Pearson product moment correlation coefficient. If there were no correlation between X and Y then

$$E(X_i - Y_i)^2 = (\mu_X - \mu_Y)^2 + \sigma_X^2 + \sigma_Y^2 \tag{3.7}$$

Krippendorff (1970, 1980) and Lin (1989) have suggested the use of an index of concordance of the following form:

$$\rho_c = 1 - \frac{\text{Observed discordance}}{\text{Chance discordance}} \tag{3.8}$$

where chance discordance is that expected if the two measurements have zero correlation. This index is analogous to Cohen's kappa statistic for the assessment of agreement between categorical ratings (see Chapter 5) and has many similarities with various intra-class correlation coefficients (see Chapters 2 and 5, and also Nickerson, 1997). Substituting (3.6) and (3.7) into Equation (3.8) yields

$$\rho_c = 1 - \frac{(\mu_X - \mu_Y)^2 + (\sigma_X^2 + \sigma_Y^2 - 2\sigma_{XY})}{(\mu_X - \mu_Y)^2 + \sigma_X^2 + \sigma_Y^2}$$

This simplifies to

$$\rho_c = \frac{2\sigma_{XY}}{(\mu_X - \mu_Y)^2 + \sigma_X^2 + \sigma_Y^2} \tag{3.9}$$

Now, if X and Y are what Tan and Inglewicz (1999) refer to as being individually equivalent ($\lambda = 1, \beta = 1, \alpha = 0$ – that is, parallel measures) then

$$\rho_c = \frac{\sigma_{XY}}{\sigma_X^2} = \frac{\sigma_{XY}}{\sigma_X \sigma_Y} = \rho \tag{3.10}$$

If we consider concordance between repeats of the measure X (say, X and X'), then

$$\rho_c = \frac{\sigma_{XX'}}{\sigma_X^2} \tag{3.11}$$

This is the reliability ratio for the standard measure, X (see Section 1.7).

The Krippendorff–Lin concordance correlation is estimated by

$$\hat{\rho}_c = \frac{2S_{XY}}{(\bar{X} - \bar{Y})^2 + S_{XX} + S_{YY}} \tag{3.12}$$

where S_{XY} is the sample estimate of the covariance between X and Y, and S_{XX} (or S_X^2) and S_{YY} (or S_Y^2) are the corresponding variances. We will usually use the denominator $n - 1$ in the estimation of variances and covariances, but if we were to use n instead then they would be maximum likelihood.

Lin (1989) showed that $\hat{\rho}_c$ is asymptotically normally distributed with mean ρ_c and variance

$$\sigma_{\hat{\rho}_c}^2 = \frac{1}{n-2} \left[\frac{(1 - \rho^2)\rho_c^2(1 - \rho_c^2)}{\rho^2} + \frac{4\rho_c^2(1 - \rho_c)u^2}{\rho} - \frac{2\rho_c^4 u^4}{\rho^2} \right] \tag{3.13}$$

where $u = (\mu_X - \mu_Y)/\sqrt{(\sigma_X \sigma_Y)}$.

Lin (1989) also demonstrated that we can improve on this normal approximation by using the inverse hyperbolic tangent transformation (or Fisher's Z-transformation),

$$\hat{Z} = \tan h^{-1}(\hat{\rho}_c) = \frac{1}{2} \ln \frac{1 + \hat{\rho}_c}{1 - \hat{\rho}_c} \tag{3.14}$$

This yields better asymptotic normality with mean $Z = \frac{1}{2}\ln[(1 + \rho_c)/(1 - \rho_c)]$ and variance

$$\sigma_Z^2 = \frac{1}{n-2} \left[\frac{(1 - \rho^2)\rho_c^2}{(1 - \rho_c^2)\rho^2} + \frac{4\rho_c^2(1 - \rho_c)u^2}{\rho(1 - \rho_c^2)^2} - \frac{2\rho_c^4 u^4}{\rho^2(1 - \rho_c^2)^2} \right] \tag{3.15}$$

The components of (3.13) and (3.15) may be replaced by their sample estimates to obtain confidence intervals and to carry out significance testing. Weighted versions and generalized forms of the concordance correlation coefficient have been described by Chinchilli et al. (1996) and by King and Chinchilli (2001), respectively, illustrating the equivalence to Cohen's weighted and simple kappa coefficients for categorical or ordinal ratings (see Chapter 5).

Although it is motivated in a completely different way the intra-class correlation (ρ_i) estimated via two-way random effects ANOVA (see Chapter 2) has a very similar form to the

Krippendorff–Lin concordance correlation. One useful computational formula for $\hat{\rho}_i$ is

$$\hat{\rho}_i = \frac{S_t^2 - S_d^2}{S_t^2 + S_d^2 + \frac{2}{n}\left(n\bar{d}^2 - S_d^2\right)} \tag{3.16}$$

where S_t^2 and S_d^2 are the sample variances of $t = X + Y$ and $d = X - Y$, respectively, and $\bar{d} = \bar{X} - \bar{Y}$ (Armitage *et al.*, 2002). A little algebraic manipulation shows that this expression is identical to

$$\hat{\rho}_i = \frac{2S_{XY}}{(\bar{X} - \bar{Y})^2 + S_X^2 + S_Y^2 - S_d^2/n} \tag{3.17}$$

The sampling properties of $\hat{\rho}_i$ have been discussed in Chapter 2. Here we simply note that the expressions for $\hat{\rho}_i$ (3.17) and $\hat{\rho}_c$ (3.12) differ only by the term $-S_d^2/n$ in the denominator of $\hat{\rho}_i$. In practice, with reasonably large sample size they will be more or less identical (Nickerson, 1997).

The major problem with these indices of concordance lies in their interpretation. What value indicates good concordance? What value indicates concordance that is good enough? The answer depends on the field of application. The use of these indices for the string data in Table 1.2, for example, would produce values well above 0.90, but this does not indicate that guessing is a good method of length measurement – unless we have the very modest aim of discriminating between the bits of string. A second problem (or at least pitfall for the unwary) is that they, like correlation coefficients in general, are dependent on the heterogeneity of the population being measured. Great care should be taken in the comparison of concordance coefficients obtained from different samples of subjects. These coefficients should be considered, along with other summary statistics, as just one more indicator of the way X and Y co-vary. The comparison of the estimate of the product moment correlation (ρ) with that obtained for either ρ_c or ρ_i, for example, is an indicator of the presence of systematic biases (or scale differences). If they are the same (or pretty close) we are primarily concerned with random measurement errors; if ρ_c or ρ_i are markedly lower than ρ then there are systematic effects to be explored. Consider the data given in Table 3.3 (kitchen scales A and B), Table 3.5 (HPLC and FPIA measures of CsA) and Table 3.6 (turbine and displacement measures of fuel consumption). The estimates of the product moment correlation in these tables are 0.9995, 0.9312 and 0.9930, respectively. The corresponding values for ρ_c are 0.9989, 0.9230, and 0.1429. The data with the clear scaling differences (fuel consumption) stands out from the two other examples. Note that $|\rho_c| \leq |\rho_i| \leq |\rho|$.

3.4 Regression with measurement error: the identifiability problem

Consider n pairs of measurements, X_i and Y_i ($i = 1, 2, ..., n$) each one being regarded as an additive combination of 'truth' and 'error' in the following way.

$$X_i = \tau_i + \delta_i$$
$$Y_i = \alpha + \beta\tau_i + \varepsilon_i \tag{3.18}$$

where τ_i (and $\beta\tau_i$) are the true values, and δ_i and ε_i are the corresponding measurement errors.

We assume

$$E(\delta_i) = E(\varepsilon_i) = 0, \text{Var}(\delta_i) = \sigma_\delta^2 \text{ and Var}(\varepsilon_i) = \sigma_\varepsilon^2, \quad \text{for all } i,$$

$$\text{Cov}(\delta_i, \delta_j) = \text{Cov}(\varepsilon_i, \varepsilon_j) = 0, \qquad\qquad \text{for all } i \neq j$$

$$\text{Cov}(\delta_i, \varepsilon_j) = 0 \qquad\qquad \text{for all } i, j$$

We also assume that the τ_i are independently distributed random variables that are also independent of the error terms. Let $E(\tau_i) = \mu$ and $\text{Var}(\tau_i) = \sigma_\tau^2$. The model is an example of a structural model (Cheng & Van Ness, 1999) and Equations (3.18) describe a linear structural relationship between X and Y. If the X_i were not random variables but were the result of a systematic choice of values by the investigator, then we would be concerned with a linear functional relationship or functional model (Cheng & Van Ness, 1999). In this discussion we will be primarily concerned with the estimation of the parameters of a structural model, but will occasionally refer to the functional model where appropriate (but see Section 4.11).

The structural model described by (3.18) implies the following structure for the covariance matrix for X and Y:

$$\begin{bmatrix} \sigma_\tau^2 + \sigma_\delta^2 & \\ \beta\sigma_\tau^2 & \beta^2\sigma_\tau^2 + \sigma_\varepsilon^2 \end{bmatrix} \tag{3.19}$$

In addition,

$$E(X_i) = \mu$$
$$E(Y_i) = \alpha + \beta\mu \tag{3.20}$$

It might be supposed that equating these expectations with their corresponding sample statistics and solving the resulting set of simultaneous equations would produce estimates of the model parameters. Unfortunately, however, this model is not identified as there are too many parameters to be estimated compared to the number of moment terms. There are four parameters describing the structure of the covariance matrix, but only 3 variance/covariance estimates. We need to make assumptions regarding one parameter in order that others may be estimated. These assumptions may be related to one or other of the methods about which we may have some prior knowledge (usually the standard, X). Alternatively, the assumption may be related to the relationship between X and Y. The following are some of the common options for assumptions to attain identifiability:

1. The precision of the standard measurement (equivalent to σ_δ^2) is known.
   (Similarly, we could make an assumption about the precision of the new method, but this would be a bit unrealistic in the present circumstances).
2. The reliability ratio of the standard, $\kappa_x = \sigma_\tau^2/(\sigma_\tau^2 + \sigma_\delta^2)$, is known.
3. The gradient of scale of Y relative to X (i.e. β) is known. For example if the scales of measurement can be assumed to be identical (and there are no proportional biases) then β equals 1. If β is some other known constant then Y can simply be re-scaled prior to any analysis to ensure that $\beta = 1$.
4. The intercept term α is known. For example it may be reasonable to assume that α equals zero.
5. The precision ratio, $\lambda = \sigma_\varepsilon^2/\sigma_\delta^2$, is known (alternatively, we could also assume that both σ_δ^2 and σ_ε^2 are known).

Having made a convincing assumption to attain identifiability, however, we still may have problems. There is a possibility that we might obtain inadmissible estimates (that is, negative variance estimates – particularly for σ_δ^2 or σ_ε^2). Having equated observed variances and covariance with their expected values, the required estimates for σ_τ^2, σ_δ^2 and σ_ε^2 are obtained from:

$$S_{XX} = \hat{\sigma}_\tau^2 + \hat{\sigma}_\delta^2$$
$$S_{YY} = \hat{\beta}^2 \hat{\sigma}_\tau^2 + \hat{\sigma}_\varepsilon^2 \qquad (3.21)$$
$$S_{XY} = \hat{\beta}^2 \hat{\sigma}_\tau^2$$

Clearly, we must have $S_{YY} > S_{XY}$ and $S_{XX} > S_{XY}/\hat{\beta}$ for the estimates of σ_δ^2 and σ_ε^2 to be nonnegative, respectively. Assuming $\sigma_\tau^2 > 0$, we also need the following to apply in order to avoid a negative estimate for this variance:

$$S_{XX} \geq \frac{S_{XY}}{\hat{\beta}}$$
$$S_{YY} \geq \hat{\beta} S_{XY}$$
$$S_{XX} \geq \hat{\sigma}_\delta^2 \qquad (3.22)$$
$$S_{YY} \geq \hat{\sigma}_\varepsilon^2$$
$$\text{sign}(S_{XY}) = \text{sign}(\hat{\beta})$$

Negative variance estimates can arise from one or both of two sources. Either (a) the model is mis-specified (i.e. not the correct one), or (b) the sample size is insufficient. If the model is correct, however, and the sample size is large enough, then we will get admissible estimates. Several authors (see Jaech, 1985, for example; or Carter, 1981) set negative variance estimates to zero and then re-estimate the other parameters subject to this constraint (using restricted maximum likelihood, for example). Although this technical 'trick' gets round the estimation problem, it does not really solve the scientific ones. Negative variance estimates are usually acting as a warning that we have to seriously think about both our model and the suitability of the data for this type of modelling. I would not recommend setting negative variance estimates to zero without giving the problem a great deal of thought.

And, finally, lack of identifiability and the above admissibility conditions are not the end of our problems! There is no guarantee that we will be able to get a finite joint confidence interval for α and β even when we obtained a sensible point estimate (Gleser & Hwang, 1987). It appears, however, that as long as σ_τ^2 is reasonably large compared to σ_δ^2 (that is the reliability ratio, κ_X, is greater than about 0.5) we are not likely to have problems in construction of a reasonable confidence interval (Gleser, 1987; see also Cheng & Van Ness, 1999).

3.5 Properties of the standard (X) (e.g. precision or reliability) known

In this section we either assume that we know the variance of the errors for the standard method, X (i.e. σ_δ^2) or, alternatively, its reliability (κ_X). Note that they are not quite the same. The latter is an assumption concerning the relative size of σ_τ^2 and σ_δ^2, not an assumption concerning the absolute value of σ_δ^2 alone.

First, let us assume that the precision of the standard method is known (that is, we know σ_δ^2). This might be reasonable if this method had been investigated thoroughly in other precision studies and had also been in routine use for a considerable time. If we know σ_δ^2 then it can be easily shown that we can obtain the following estimates:

$$\hat{\beta} = \frac{S_{XY}}{\left(S_{XX} - \sigma_\delta^2\right)} \tag{3.23}$$

$$\hat{\sigma}_\tau^2 = S_{XX} - \sigma_\delta^2 \tag{3.24}$$

$$\hat{\sigma}_\varepsilon^2 = \frac{S_{YY} - S_{XY}^2}{\left(S_{XX} - \sigma_\delta^2\right)} \tag{3.25}$$

On the assumption of multivariate normality, these moments estimators would be maximum likelihood if we were to replace S_{XX}, S_{YY} and S_{XY} by $S_{XX}(n-1)/n$, $S_{YY}(n-1)/n$ and $S_{XY}(n-1)/n$, respectively. The asymptotic variance-covariance matrix of the maximum likelihood estimates may be obtained (i.e. estimated) via the inverse of the observed information matrix. Hood *et al.* (1999) provide closed form expressions for both the asymptotic information and variance-covariance matrices for the linear structural model as described in equation (3.18) under various assumptions necessary for identifiability. However, as we do not make use of most of these closed form expressions explicitly in this text we will not report them here. From the model parameter estimates and their variance-covariance matrix we can also obtain estimates of the precision of Y, its relative precision or reliability and also ψ. Bootstrap sampling and other methods such as Taylor expansions (the Delta technique) can be used to estimate standard errors or confidence intervals for any of these parameters or combinations of them. Again, we should bear in mind the possibility (and in the case of small samples, a fairly high probability) of negative variance estimates simply arising from sampling fluctuations.

Suppose that we know κ_X. The required estimators are:

$$\hat{\beta} = \frac{S_{XY}}{\left(\kappa_X S_{XX}\right)} \tag{3.26}$$

$$\sigma_\tau^2 = \kappa_X S_{XX} \tag{3.27}$$

$$\hat{\sigma}_\delta^2 = S_{YY} - \frac{S_{XY}^2}{\left(\kappa_X S_{XX}\right)} \tag{3.28}$$

$$\hat{\sigma}_\varepsilon^2 = S_{XX}(1 - \kappa_X) \tag{3.29}$$

Again, the standard errors of the parameter estimates may calculated via the observed information matrix.

Now, let us briefly consider the correlation between X and Y ($\rho_{XY} \equiv \rho$):

$$\rho_{xy} = \frac{\mathrm{Cov}(X_i, Y_i)}{(\mathrm{Var}(X_i)\,\mathrm{Var}(Y_i))^{1/2}}$$

or, equivalently,

$$\rho_{XY}^2 = \frac{\mathrm{Cov}(X_i, Y_i)^2}{\mathrm{Var}(X_i)\,\mathrm{Var}(Y_i)} \tag{3.30}$$

Now, it is straightforward to show that, under the assumptions of the model (3.18),

$$\rho_{XY}^2 = \kappa_X \kappa_Y \tag{3.31}$$

It follows that if κ_X were known, then the new measure, Y, would be more reliable, or more precise, than X (that is $\psi > 1$) if and only if $\rho_{XY}^2 > \kappa_X^2$. Usually we would expect ρ_{XY} to be positive and, by definition, the reliabilities must be positive and less than or equal to 1. Hence the often-repeated suggestion that the product moment correlation has no role in method comparison studies is not true.

It follows from Equation (3.31) that $\psi > 1$ if $r_{XY} > \kappa_X$, where r_{XY} is the observed product moment correlation ($\hat{\rho}_{XY}$) between X and Y. κ_Y is simply estimated by r_{XY}^2/κ_X. With κ_X assumed known we can estimate precision from confidence intervals of r_{XY}^2. Alternatively, its sampling distribution can be examined through bootstrapping. Again, we emphasise the possibility of negative variance estimates (leading to improper estimates of reliability). An alternative parameterisation of the model is provided by Shyr and Gleser (1986), who also discuss derivations of standard errors and significance tests concerning estimates of precision. They also discuss sample size and power. Their conclusion concerning the latter is that we need fairly large sample sizes if we are seriously considering tests of the relative precision of two measurement methods. The general problem, however, is whether we can ever be confident that we actually know the value of κ_X. It is usually a fallible estimate from some previous data set.

3.6 Grubbs estimators: when β assumed to be 1

One simple assumption, that is quite often justified *a priori*, is that the scale of measurement for the two instruments or assays is the same (that is, $\beta = 1$). In this case

$$\mathrm{Var}(X_i) = \sigma_\tau^2 + \sigma_\delta^2$$

$$\mathrm{Var}(Y_i) = \sigma_\tau^2 + \sigma_\varepsilon^2 \tag{3.32}$$

$$\mathrm{Cov}(X_i, Y_i) = \sigma_\tau^2$$

Equating these expected values with the corresponding sample statistics provides moment estimators of the following form:

$$\hat{\sigma}_\tau^2 = S_{XY} \tag{3.33}$$

$$\hat{\sigma}_\delta^2 = S_{XX} - S_{XY} \tag{3.34}$$

$$\hat{\sigma}_\varepsilon^2 = S_{YY} - S_{XY} \tag{3.35}$$

They were first derived by Grubbs (1948, 1973) and are therefore frequently referred to as Grubbs estimators. Under the assumption of multivariate normality, and if the sample statistics were to be calculated using the divisor n rather than the usual $n-1$ then the estimates in (3.33) to (3.35) would also be maximum likelihood. The standard errors of the Grubbs' estimators for the two measurement error variances are given by Grubbs (1948, 1973):

$$
\begin{aligned}
Var(\sigma_\delta^2) &= \frac{2\sigma_\delta^4}{n-1} + \frac{\sigma_\tau^2\sigma_\delta^2 + \sigma_\tau^2\sigma_\varepsilon^2 + \sigma_\delta^2\sigma_\varepsilon^2}{n-1} \\
Var(\sigma_\varepsilon^2) &= \frac{2\sigma_\varepsilon^4}{n-1} + \frac{\sigma_\tau^2\sigma_\delta^2 + \sigma_\tau^2\sigma_\varepsilon^2 + \sigma_\delta^2\sigma_\varepsilon^2}{n-1}
\end{aligned}
\tag{3.36}
$$

Again, we can construct approximate confidence intervals for the parameter estimates and functions of them from the observed information matrix obtained via maximum likelihood. Exact confidence intervals are given in Thompson (1963) and Jaech (1985).

Based on work of Pitman (1939) and Morgan (1939), a simple significance test for the difference between the two error variances has been described by Maloney and Rastogi (1970). If $d = X - Y$ and $s = X + Y$ then the expected covariance of d and s is zero if and only if the two error variances are equal, equivalent to ψ equal to 1 in this case. One therefore constructs a test statistic

$$
t = \frac{r_{ds}(n-2)^{1/2}}{(1 - r_{ds}^2)^{1/2}}
\tag{3.37}
$$

where r_{ds} is the observed product moment correlation between d and s. Under the null hypothesis ($\sigma_\delta^2 = \sigma_\varepsilon^2$) t is distributed as Student's t with $n-2$ degrees of freedom. There is no need to calculate u and v explicitly, but note that

$$
r_{ds} = \frac{S_{ds}}{(S_{dd}S_{ss})^{1/2}}
$$

where

$$
\begin{aligned}
S_{dd} &= S_{XX} + S_{YY} - 2S_{XY} \\
S_{ss} &= S_{XX} + S_{YY} + 2S_{XY} \\
S_{ds} &= S_{XX} - S_{YY}
\end{aligned}
$$

As the calculations for the Maloney and Rastogi test are so straightforward I will not illustrate them using an example at this point. It is, however, worth having a look at the potential power of the test. Table 3.8 shows the results of a series of Monte Carlo simulation experiments in which the test statistic described in Equation (3.37) was used to compare the error variances of two measuring instruments (see Dunn, 1992). Under each condition (a specified value of σ_τ^2) a thousand sets of measurements were generated and the test statistic calculated. The power of the test was estimated from the proportion of the thousand test statistics that were significant at the 0.05 level (a 1-tailed test of the hypothesis that Y is more precise than X, i.e. $\psi < 1$). In all the simulations, $\sigma_\delta^2 = 16$ and $\sigma_\varepsilon^2 = 4$ (i.e. $\psi = 0.25$). Incidentally, this is equivalent to testing the difference between κ_x and κ_y based on this

Table 3.8 Results of simulated method comparison studies using the Maloney and Rastogi test.

Sample size	σ_τ^2	κ_x^*	κ_y^*	Estimated power
25	0	–	–	0.95
25	1	0.06	0.20	0.91
25	4	0.20	0.50	0.74
25	25	0.61	0.86	0.32
25	100	0.86	0.96	0.16
25	400	0.96	0.99	0.08
25	1600	0.99	0.998	0.06
50	100	0.86	0.96	0.23
100	100	0.86	0.96	0.24

* Reliability ratio.

simple paired comparison. It can be seen that if there is considerable variation in the items being measured (relative to that of the measurement errors) the power of the comparison is very low. As one would frequently be dealing with a wide variation in the items being measured in a study of this sort, this result illustrates that the design of the study (that is, paired measurements without replication) is inadequate. It is quite straightforward to use bootstrap or the Delta technique to provide a confidence interval for ψ (i.e. $\sigma_\varepsilon^2/\sigma_\delta^2$) but, again this would be rather wide for anything except rather large samples. Formal calculations of sample sizes for testing the equality of the precisions of two instruments using this design can be found in Shoukri and Demirkaya (2000).

We might wish to simultaneously test whether $\alpha = 0$ and $\psi = 1$ (that is, individual equivalence). This can be done using a procedure described by Bradley and Blackwood (1989), again an extension of the work of Pitman (1939) and Morgan (1939). First, let

$$E(s) = \mu_X + \mu_Y$$
$$E(d) = \mu_X - \mu_Y$$

Then

$$E(d \mid s) = (\mu_X - \mu_Y) + \rho_{sd}\left(\frac{\sigma_d}{\sigma_s}\right)(s - (\mu_X - \mu_Y))$$

$$= (\mu_X - \mu_Y) + \left[\frac{\sigma_X^2 - \sigma_Y^2}{\sigma_s^2}\right][s - (\mu_X - \mu_Y)]$$

$$= \alpha^* + \beta^* s$$

where

$$\alpha^* = (\mu_X - \mu_Y) - \left[\frac{\sigma_X^2 - \sigma_Y^2}{\sigma_s^2}\right](\mu_X + \mu_Y)$$

and

$$\beta^* = \frac{\sigma_X^2 - \sigma_Y^2}{\sigma_s^2}$$

Table 3.9 Grubbs estimators: simulated data (from Dunn & Roberts, 1999).

Sample size	True values			No. inadmissible* estimates of σ_δ^2 (out of 100 simulated sets of data)
	σ_τ^2	σ_δ^2	σ_ε^2	
100	100	10	10	1
100	100	50	10	11
100	100	10	1	45
100	25	10	1	26
100	5	10	1	10

* Less than or equal to zero.

Now $\sigma_X^2 = \sigma_Y^2$ (that is, $\sigma_\delta^2 = \sigma_\varepsilon^2$) and $\mu_X = \mu_Y$ (that is, $\alpha = 0$) if and only if $\alpha^* = \beta^* = 0$. So, a simultaneous test of $\alpha = 0$ and $\psi = 1$ is an F-test calculated from the standard results of a regression of the difference (d) on the sum (s). The test statistic is

$$F = \frac{\left[\left(\sum_i d_i^2 - SSE \right) / 2 \right]}{SSE/(n-2)} \tag{3.38}$$

with 2 and $n - 2$ degrees of freedom, and where SSE is the sum of squares of the residuals from a regression of d on s (see also Hawkins, 2002).

Now let's turn to some data and simply calculate the Grubbs' estimates of the two measurement error variances. Starting with the data using the kitchen scales (Table 3.2) we can immediately see from inspection of the covariance matrix in Table 3.3 that the estimated variance for Scale B is less than the estimated covariance for A and B. We have a negative variance estimate for the measurement errors for Scale B! Moving on to the kidney concentrations for CsA (Tables 3.4 and 3.5) we can estimate the variance of the HPLC from $64522.2 - 63143.2 = 1379$. Similarly, the measurement error variance for FPIA = $71264.1 - 63143.2 = 8120.9$. $\hat{\psi} = 8120.9/1379 = 5.889$. If, however, we were unhappy with the possible outlier in Table 3.4 and therefore delete this point from the data, we then run into a negative variance estimate again. Even Grubb's original data (Table 3.1) yields some negative variance estimates! I will not bother with any attempts at formal inference (standard errors or confidence intervals) for any of these estimates.

Table 3.9 further illustrates the problem of negative variance estimates and its dependence on sample size. Here we have used Monte Carlo simulation using various combinations of sample size and values of σ_τ^2, σ_δ^2 and σ_ε^2. The message is clear. Again we have to conclude that, *even* if we have the correct model, the estimation of the instruments' precisions, and especially their comparison, is only viable when we have large sample sizes. But how often can we assume with confidence that β is *exactly* equal to 1?

3.7 Fitting a line through the origin ($\alpha = 0$)

There might often be situations where we might expect that the intercept term is equal to zero. But, as in the last section, can we be sure *a priori* that it will be exactly zero? The assumption is probably unsafe, but for completeness, I give the estimates for the other parameters given this constraint, and then move on. They are

$$\hat{\beta} = \frac{\overline{Y}}{\overline{X}}$$

$$\hat{\sigma}_\tau^2 = \frac{S_{XY}}{\hat{\beta}}$$

$$\hat{\sigma}_\delta^2 = S_{XX} - \frac{S_{XY}}{\hat{\beta}}$$

$$\hat{\sigma}_\varepsilon^2 = S_{YY} - \hat{\beta}S_{XY}$$

(3.39)

3.8 'Orthogonal' regression: $\lambda(=\sigma_\varepsilon^2/\sigma_\delta^2)$ assumed to be known

For a given λ, and assuming the joint normality of τ, δ and ε, the moments estimate (also maximum likelihood if S_{XX}, etc. are the ML estimates) of β is given by:

$$\hat{\beta} = \frac{S_{YY} - \lambda S_{XX} + \left\{(S_{YY} - \lambda S_{XX})^2 + 4\lambda S_{XY}^2\right\}^{1/2}}{2S_{XY}}$$

(3.40)

α is simply estimated by

$$\hat{\alpha} = \overline{Y} - \hat{\beta}\overline{X}$$

(3.41)

The above estimates were first derived by Kummel (1879) but are usually attributed to Deming (1943) in the clinical chemistry literature (hence the commonly-used term 'Deming's regression'). Here we use 'orthogonal regression' but strictly speaking it is only an orthogonal regression when $\lambda = 1$ (see discussion of least squares at the end of this present section). Following from the above estimate of β it can be shown that the corresponding estimates of σ_τ^2 and σ_ε^2 are the following, respectively:

$$\hat{\sigma}_\tau^2 = \frac{S_{XY}}{\hat{\beta}}$$

$$\hat{\sigma}_\delta^2 = \frac{S_{YY} - 2\hat{\beta}S_{XY} + \hat{\beta}^2 S_{XX}}{\lambda + \hat{\beta}^2}$$

(3.42)

In this situation the investigators are usually concerned with the sampling characteristics of $\hat{\beta}$ but have little if any interest in the other parameters. Tan & Iglewicz (1999), for example, were concerned with confidence intervals for $\hat{\beta}$ or $\hat{\beta}/\sqrt{\lambda}$ (the latter being equivalent to $1/\sqrt{\hat{\psi}}$. A corresponding confidence interval for $\hat{\psi}$, itself, could also be constructed. Exact confidence intervals for $\hat{\beta}$ can be constructed using the methods of Creasy (1956) and Williams (1959, 1973). Tan & Iglewicz (1999) use these methods for this particular application, and they are also described by Cheng & Van Ness (1999). Here, however, we will use large sample approximations based on the inverse of the observed information matrix (see, for

Table 3.10 Summary statistics for Strike's gentamicin data ($n = 56$).

	EMIT1	EMIT2	FIA1	FIA2
Mean	5.9518	5.5946	5.8393	5.8750
S.D.	4.8054	4.2906	4.2111	4.2158
Covariance matrix				
	23.3972			
	20.2946	18.7267		
	19.4823	17.1864	17.3465	
	19.7908	17.5527	17.5766	17.9724
Correlations				
	1	1		
	0.9560	0.9409	1	
	0.9652	0.9557	0.9889	1

example Hood *et al.*, 1999; Cheng & Van Ness, 1999). It can be shown that $\hat{\beta}$ is asymptotically normally distributed with variance given by

$$\sigma_\beta^2 = \frac{\sigma_\delta^2 \sigma_\varepsilon^2 + \beta^2 \sigma_\tau^2 \sigma_\delta^2 + \sigma_\tau^2 \sigma_\varepsilon^2}{n \sigma_\tau^4} \tag{3.43}$$

This variance is estimated by substitution of the estimates of its components into (3.43).

Strike (1991, 1995) illustrated the method using two assays for gentamicin. The performance of a new assay (FIA) is compared to that of the reference assay (EMIT). Each assay is used twice on each of 56 specimens. Strike *estimated* λ using the replicates and then used the Deming method to estimate β and its standard error using the mean of the FIA replicates as Y and the mean of the EMIT replicates as the corresponding X. Summary statistics for these data are given in Table 3.10 (the raw data can be found in Strike's two books). We discuss Strike's gentamicin data in further detail in Chapter 4, but in the meantime simply report his estimate of β to be 0.950 with an approximate standard error of 0.035. Strike's estimates for α, σ_ε^2, and σ_δ^2, were 0.373, 0.823 and 0.194, respectively (the latter two leading to an estimated λ of 0.236). The estimate of ψ is 3.87.

The crucial problem we are faced with when using the Deming regression method is whether we can be sure that we really know the value of λ. Lakshminarayanan and Gunst (1984), Lewis *et al.* (1991) and Linnet (1998) have investigated the sensitivity of the estimates of β to various choices of λ, but this does not really help in the present application. Even if we ignore the fact that we are often estimating λ from concurrent or previous data on replications of the measurements, rather than really using a value known *a priori*, can we be sure that we are actually estimating the correct ratio? This choice probably has more influence on the value of ψ (i.e. λ/β^2) than does the sensitivity of the estimates of β.

Carroll, Ruppert and Stefanski (1995) and Carroll and Ruppert (1996) have rightly been very critical of much of the naïve use of this methodology. Typically, λ is estimated from repeatability standard deviations (or variances) – as in the case of the Strike example, above. This, these critics refer to as 'duplicity'. It might, however, be more appropriate to use reproducibility standard deviations, instead. Mandel (1991), for example, illustrated the Deming method using an example involving numerical (X) and subjective (Y) assessments of the quality of dried eucalypti veneers (using data from Kauman *et al.*, 1956). For each of the two methods, each veneer was assessed twice by each of three raters (the standard deviation of the six assessments was then taken as an estimate of the reproducibility standard deviation to be used in the calculation of λ). This is certainly an improvement on the usual procedure of using simple replications but, again, it has some deficiencies. If there are subject- or item-specific biases (matrix effects – see Sections 1.4, 3.9 and 4.10) in the use of either or both of the measurement methods then there is at least one random component of the measurement error that is being missed by simple replication. If we simply repeat the measurements using one of the methods, the item-specific biases will be completely confounded with the true values. The variance of the measurement errors will be underestimated. Mandel's approach in the veneer example might have picked up this source of variability, but we cannot be sure (at least, not on the basis of the analysis which he presented). Carroll and Ruppert (1996) give an example of a simpler case in which there is, say, a single measurement of X (but with a known measurement error variance), but that of Y is measured once and then measured again some time later (to give Y_1 and Y_2, say). An estimate σ_ε^2 is obtained using the two measures of Y and then used to calculate λ for the subsequent orthogonal regression of $(Y_1 + Y_2)/2$ against X. In the presence of a systematic shift of the values of Y between the first and second measurements they point out that σ_ε^2 should be estimated by

$$\hat{\sigma}_\varepsilon^2 = \frac{\sum_i \left\{ (Y_{i2} - Y_{i1}) - (\overline{Y}_2 - \overline{Y}_1) \right\}^2}{4(n-1)} \tag{3.44}$$

rather than by

$$\hat{\sigma}_\varepsilon^2 = \frac{\sum_i (Y_{i2} - Y_{i1})^2}{4n} \tag{3.45}$$

which is the usual choice based on repeatability. We illustrate this problem in greater detail in Section 3.9.

Although, until now, we have approached the problems of regression of Y on X when both variables are subject to error, in terms of equating observed variance-covariance matrices to their expected (modelled) values, an equivalent, and historically more common, approach would be to think of modifying the least squares method as used in the more common use of classical linear regression. Here we restrict the discussion to the case when λ is supposedly known (a more general discussion can be found in Cheng & Van Ness, 1999).

Classical least squares minimises the sum of squares of the vertical distances of the observations from the fitted line, because the Ys are subject to error. Measurement error models, however, have to take account of the measurement errors in *both* the Xs and the Ys. Because the errors of measurement in X and Y are assumed to be independent, we minimise

$$SS = \sum_i \frac{(X_i - \tau_i)^2}{\sigma_\delta^2} + \sum_i \frac{(Y_i - \alpha - \beta\tau_i)^2}{\sigma_\varepsilon^2} \tag{3.46}$$

The resulting estimators are exactly the same as those already derived, above. But the use of equation (3.46) naturally leads to some possible modifications. If λ is fixed, but the size of, say, σ_δ^2, varies with the size of the characteristic being measured (τ), then we can use a weighted regression procedure with weights, for example, inversely proportional to our estimated values of τ or τ^2. Linnet (1990), for example, uses iterative weights (w_i) that are given by the following:

$$w_i = \frac{1}{[(\hat{X}_i + \hat{Y}_i)/2]^2} \qquad (3.47)$$

where the starting values for $\hat{X}_i$ and $\hat{Y}_i$ are the observed measurements. Other weighting methods are described by Martin (2000). If it is known that λ is not fixed but varies with the size of the characteristic being measured (τ), then the methods proposed by Nix and Dunstan (1991) are more appropriate.

Returning to equation (3.46), if we assume that the variances of the errors in X and Y are equal (i.e. $\lambda = 1$) then this is equivalent to fitting a line by minimising the sum of the squared perpendicular distances between the observations and the fitted line (hence 'orthogonal' regression). This was the basis of the first work in this area (Adcock, 1877, 1878) and of Pearson's derivation of what are now known as principal components (Pearson, 1901). Kummell (1879) was the first to deal with the more general situation in which λ can take any known value (see Finney, 1996, for a brief history).

3.9 Equation error

Let us now assume that the relationship between X and Y is described by the following:

$$X_i = \tau_i + \delta_i$$
$$Y_i = \alpha + \beta\tau_i + \eta_i + \varepsilon_i \qquad (3.48)$$

where the new term η_i is a random variable with mean of zero and variance σ_η^2. η_i is also assumed to be independent of τ_j, δ_j and ε_j for all i and j. The new random component is not necessarily a measurement error but is a part of Y that is not related to the construct or characteristic being measured (i.e. τ). Unlike ε_i, η_i does not vary with replications (i.e. it is stable). Mathematical statisticians call it equation error (Fuller, 1987; Carroll, Ruppert & Stefanski, 1995; Cheng & Van Ness, 1999). But how should equation error be understood in the present context? It has already been introduced as a component of specificity in the psychometricians' factor analysis model (see Section 1.4). It could also be thought of as an indicator of invalidity: if Y were perfectly valid (with respect to X) its true value (here $\beta\tau + \eta$) would be perfectly correlated with that of X (i.e. τ). Perfect correlation cannot be achieved if $\sigma_\eta^2 > 0$.

In general, the item or subject being measured invariably has some property or characteristic, other than that we are trying to measure, which 'interferes' with the measurement process. The amount of interference will be dependent on the measuring instrument or method under consideration. This interference or non-specificity is often referred to as item-specific bias. It appears to have been first discussed by Fairfield Smith (1950) and is well-documented in the literature of analytical chemistry (Mandel, 1964) and clinical chemistry (Lawton et al., 1979; Strike, 1991, 1995). It was discussed by Cochran (1968), who referred to an unpublished manuscript on an inter-rater reliability study by Mosteller. In clinical chemistry it is frequently referred to as a random matrix effect, the matrix being the tissue or bodily fluid containing the material to be measured. Blood serum, for example, may contain materials other than the one under investigation that interfere with or affect the results of a particular type of assay procedure. From a statistical point of view, its main characteristic is that it cannot be detected from unreplicated measurements. In my experience, however, it is certainly not a problem that is solely a characteristic of chemical measurements. It appears to be displayed in the lung-function data described by Barnett (1969, 1987). It is also illustrated by both the Computerized Axial Tomography (CAT) scan and EEG data discussed by Dunn (1989, 1992). The CAT scan data will be used for illustrative purposes in this section. The analysis of several sets of EEG data is given in Besag *et al.* (1989). In the context of ANOVA models for interviewer or rater reliability studies, the above item-specific biases are equivalent to interviewer-subject or rater-subject interactions (Fleiss, 1970; Shrout & Fleiss, 1979).

> In our experience, outside of some special laboratory validation studies, equation error is almost always important in linear regression.
>
> (Carroll, Ruppert & Stephanski, 1995)

> When its assumptions hold, orthogonal regression is a perfectly justifiable method of estimation. As a method, however, it often lends itself to misuse by the unwary because orthogonal regression does not take equation error into account.
>
> (Carroll & Ruppert, 1996)

With these warnings in mind, we will now explain the implications of the presence of equation error in equation (3.48). If there were to be equation error for the Xs life would be even more complicated! The more complex situation will be left for the following chapter and we will here naively assume that (3.48) is, in fact, the correctly specified model.

The expected covariance matrix implied by (3.48) is

$$\begin{bmatrix} \sigma_\tau^2 + \sigma_\delta^2 \\ \beta\sigma_\tau^2 & \beta^2\sigma_\tau^2 + \sigma_\eta^2 + \sigma_\varepsilon^2 \end{bmatrix} \tag{3.49}$$

We assume for the time being that we do not have access to replicated measures. In principle, we can still use Grubbs' estimators of the variance components as long as we are sure that $\beta = 1$. We will not be able to separate σ_η^2 from σ_ε^2, however, as they will be completely confounded. That is not necessarily a problem unless we wish to fully understand the nature of the measurement process. If σ_δ^2 is known, all of the parameters are identified except that, again, we are not able to separate σ_η^2 from σ_ε^2. The main problem arises in the valid use of orthogonal regression. Let's assume that we have access to estimates of both σ_δ^2 and σ_ε^2 from replications of the present measurements or from previous repeatability studies. This allows us to calculate $\lambda(=\sigma_\varepsilon^2/\sigma_\delta^2)$. It is fairly straightforward to show that, unfortunately, a knowledge of λ does not make (3.48) identifiable. What we really need is a knowledge of $\lambda' = \sigma_\varepsilon^2/(\sigma_\eta^2 + \sigma_\varepsilon^2)$. If we use the invalid ratio $(\sigma_\varepsilon^2/\sigma_\delta^2)$ we will be able to get estimates of β and the other parameters, but they too will be invalid. They will be biased. This is why Carroll & Ruppert (1996) are able to claim that many examples of orthogonal regression are flawed. Edland (1996) has also discussed this problem in some detail.

Let's have a look at an example. Table 3.11 provides measurements derived from CAT scans of the heads of 50 psychiatric patients (Turner *et al.*, 1986; see also Dunn, 1989, 1992). The primary aim of these scans was to determine the size of the brain ventricle relative to that of the patient's skull (the ventricle-brain ratio or VBR is equal to {(ventricle size/brain size) $\times$ 100}.

Table 3.11 CAT Scan log(VBR) data.

Plan1	Plan3	Pix1	Pix3	Plan1	Plan3	Pix1	Pix3
2.05	2.13	1.79	1.77	1.72	1.28	0.00	0.00
1.93	1.79	1.53	1.55	2.16	1.96	1.57	1.57
2.27	1.95	1.65	1.70	2.53	2.17	2.05	2.12
1.79	1.67	1.59	1.65	1.87	1.48	1.03	1.03
1.57	1.57	0.69	0.74	1.39	1.39	1.69	1.79
1.89	1.84	1.50	1.55	2.39	2.26	1.74	1.72
1.67	1.72	1.50	1.63	1.57	1.39	0.74	0.74
2.30	2.25	1.67	1.69	2.03	1.93	1.61	1.59
1.19	1.70	1.03	0.99	1.13	0.41	0.88	0.96
1.63	1.22	1.25	1.28	1.93	2.03	1.79	1.77
1.89	1.50	1.84	1.89	1.63	2.03	1.22	1.22
1.70	1.96	1.90	1.99	2.82	2.84	2.91	2.93
0.53	0.99	1.19	1.10	2.24	2.03	2.33	2.37
1.63	1.76	1.22	1.19	1.55	1.53	1.63	1.39
2.12	2.30	1.87	1.84	1.63	1.34	1.19	1.10
1.46	0.96	0.34	0.34	1.87	1.41	1.19	1.25
1.79	1.84	1.53	1.53	2.33	1.84	1.63	1.65
1.39	1.16	0.83	0.88	1.96	1.53	1.10	1.10
2.40	2.30	1.76	1.76	2.09	1.89	1.41	1.44
1.39	1.41	0.92	0.96	2.22	1.89	1.63	1.65
1.67	1.34	0.74	0.79	2.03	1.46	0.74	0.79
2.26	2.12	1.36	1.36	1.69	1.63	1.28	1.31
2.30	2.50	2.30	2.29	2.01	1.50	1.39	1.34
1.57	1.59	1.16	1.16	2.08	1.55	0.69	0.69
2.13	2.09	1.95	1.95	1.69	1.13	1.57	1.57

For a given scan or 'slice' the VBR was determined from measurements of the perimeter of the patient's ventricle together with the perimeter of the inner surface of the skull. The measurements were made using (a) a hand-held planimeter on a projection of the x-ray image, or (b) from an automated pixel count based on the image on a VDU. Table 3.11 displays the logarithms of the VBRs for single scans from the 50 patients. Like many method comparison studies, the collection of the data actually involved replication. The first two columns correspond to the repeated determinations based on pixel counts and the second two columns correspond to repeated determinations based on the use of a planimeter. Here the planimeter is regarded as the standard method (X) whilst the pixel count is the new one (Y). Based on the nature of the measurements, we would expect β to be fairly close to 1.

A much more detailed analysis of these data will be presented in Chapter 4. Here we consider the regression of the mean of the two pixel-based measurements on the mean of the two based on planimetry. We start, however, with a Bland–Altman plot (Figure 3.12) and by simply estimating the variance of the differences between the two (as in Bland and Altman's approach to the estimation of limits of agreement). The estimate is 0.1392. Again, we use the means of the replicates, $S_{XX} = 0.1575$, $S_{YY} = 0.2732$ and $S_{XY} = 0.1459$. Assuming $\beta = 1$, the use of Grubbs' methods yields ML estimates for the error variances of $0.1575 - 0.1459 = 0.012$ and $0.2732 - 0.1459 = 0.127$ for planimetry- and pixel-based measures, respectively. Note that the sum of these two estimates is 0.139 (the variance of the differences between them). So far, so good. Calculation of the correlation between the difference (Pixel − Planimetry) and the sum (Pixel + Planimetry) of the measurements yields a value of $+0.37$ ($t = 2.72$ with 48 d.f.). It appears that planimetry is statistically significantly more precise than using automated pixel counts.

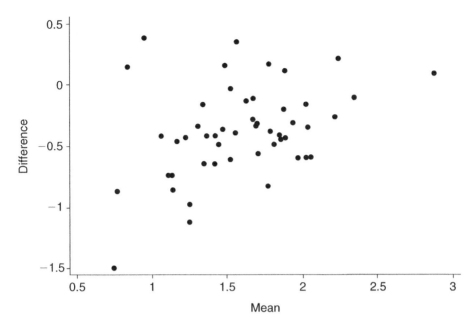

Figure 3.12   Log(VBR): Difference (Pixel − Planimeter) versus the mean, (Pixel + Planimeter)/2.

Now let's consider the estimation of β. Ignoring any possible drift in the measurements from time 1 to time 2, the means of the squared differences between time 1 and time 2 of the pixel and planimeter measures are approximately 0.002 and 0.049, respectively. Now the pixel measures appear to be far more precise than the use of a planimeter! These estimate the repeatability variances (σ_ε^2 and σ_δ^2, respectively). So, we calculate $\hat{\lambda} = 0.002/0.049 = 0.041$ (or 1/24.5). Use of orthogonal regression (i.e. equation 2.40) yields $\hat{\beta} = 1.85$. This, as we have already pointed out, is an invalid estimate, and it is not surprising that it appears to be inconsistent with our expectations.

If, instead of assuming that we know λ, we merely assume that we know the characteristics of the planimeter (i.e. assume σ_δ^2 to be $0.049/2 = 0.025$, where we divide by two because we are using the mean of two replicates in our regression analyses), then we can use Equation (3.23) to estimate β. Here $\hat{\beta} = 0.1459/(0.1575 - 0.025) = 1.10$, a value close to that expected (i.e. $\beta = 1$).

From equation (2.25) it follows that $\hat{\sigma}_\varepsilon^2 = 0.2732 - 0.1459^2/(0.1575 - 0.025) = 0.1125$. This is fairly close to the corresponding Grubbs' estimator. We can get an estimate of the equation error variance (σ_η^2) by subtracting 0.001 (half the estimated pixel repeatability variance) from 0.1125, to give a value of 0.1115.

3.10 Software for covariance structure modelling

Software for covariance structure modelling is widely available (see Appendix 6) and I shall illustrate its use through a variety of examples using Peter Bentler's EQS program (Bentler, 1995). It is not my purpose to provide details of how to use this or any other package. This would clearly be beyond the scope of this text. Instead, I provide a series of program listings to fit various models to provide an explicit reference to how the data sets have been analysed (Appendix 1). Users of other software packages should have no difficulty in translating the program listings for their own use. Readers wishing to learn how to use this package are referred to Bentler (1995) and to Dunn *et al.* (1993). We discuss general approaches to covariance structure modelling in Chapter 4. Appendices 2 to 5 will also illustrate the use of software packages for the required analyses.

We start here by fitting a simple model to the CAT scan data using 'orthogonal' regression. This is equivalent to using maximum likelihood (assuming multivariate normality) subject to the linear constraint $\lambda \sigma_\delta^2 - \sigma_\varepsilon^2 = 0$. The program listing is given in Appendix 1(a). Running this program yields an estimate of β equal to 1.852 (s.e. 0.268). $\hat{\sigma}_\delta^2 = 0.079$ (s.e. 0.016) and $\hat{\sigma}_\varepsilon^2 = 0.003$ (s.e. 0.001). The model is just identified (so the reported goodness-of-fit chi-square = 0 with 0 d.f.).

We now fit a similar model, but assuming that $\sigma_\delta^2 = 0.025$ (see Appendix 1(b)). Again the model is just identified and therefore the chi-square goodness-of-fit statistic is 0 with 0 d.f. $\hat{\beta} = 1.101$ (s.e. 0.165) and $\hat{\sigma}_\varepsilon^2 = 0.113$ (s.e. 0.029). Now, the use of the modelling software enables us to introduce and test further constraints. The obvious one here is to set $\beta = 1$, still assuming that $\hat{\sigma}_\delta^2 = 0.025$ (see Appendix 1(c)). The resulting chi-square is 0.377 with 1 d.f. (P = 0.539), indicating that β is not significantly different from unity. The revised estimate for σ_ε^2 is 0.118 (s.e. 0.028). The programme listing for Grubbs' estimators (constraining $\beta = 1$, but making no assumptions concerning the error variances) is shown in Appendix 1(d). Again, we are back to a just identified model (chi-square = 0). The required estimates are: $\hat{\sigma}_\delta^2 = 0.012$ (s.e. 0.021) and $\hat{\sigma}_\varepsilon^2 = 0.127$ (s.e. 0.033). Finally, we introduce an equality constraint in the latter program to test whether $\sigma_\delta^2 = \sigma_\varepsilon^2$ (see Appendix 1(e)). The resulting chi-square is 7.013 with 1 d.f. (P = 0.008), which is consistent with the results of the Maloney–Rastogi test in Section 2.9.

3.11 Conclusions and recommendations

The first conclusion that we must come to is that if an investigator is seriously interested in comparing the measurement characteristics of measurement instruments, then these simple designs in which n subjects or items are each measured once only using just two methods (i.e. without replication) are virtually wholly inadequate. The resulting data simply do not contain enough relevant information and no method of statistical analysis of the data will be able to reveal what we are looking for. The basic measurement model (Equation (3.18)) is not identified, and we cannot make progress without having to make assumptions that can often be subject to serious and valid criticism. And we often need to make assumptions about the very characteristics we wish to examine. Readers should always remember that plotting the difference between the two measurements against their mean (the Bland–Altman plot), however useful the plot may be, does not solve the problem. There is no more information in this plot than in an X–Y scatter plot. The underlying model is still unidentified!

If, however, the investigator has the much more modest aim of checking whether there is reasonable agreement between measurements provided by two instruments, then the Bland–Altman plot, together with estimates of limits of agreement, can answer this basic question. But the new instrument may be considerably better (in terms of its precision) than the established standard, however, and if we were only to use these simple exploratory methods we would never know that the standard was the poorer of the two! If a new thermometer, for example, happened to use a different scale of measurement (Celsius rather than the older Fahrenheit, say) we might reject it after looking at limits of agreement, even though it still might be the better (more precise) instrument. Readers might object that this problem could easily have been foreseen. This is indeed true, but differences in scaling are often much more subtle than this and should not be missed. That is, they should not be dismissed as measurement errors.

Very often the design is not the simple one described above. Frequently the design actually involves replication of the measurements taken by each of the two instruments. Bland and Altman (1986), for example, present data for the comparison of two methods of measuring peak expiratory flow rate – the Wright peak flow meter and the Mini Wright peak flow meter. Each method is used twice on each of the subjects. The authors comment:

> Only the first measurement by each method is used to illustrate the comparison of methods, the second measurements being used in the study of repeatability.

Typically, however, these authors recommend comparing the two methods using the means of the replications (see, for example, Bland & Altman, 1999). Strike (1991, 1995) does the same, but in this case uses the means as the source of data for an orthogonal regression. I have also done this in the present chapter to illustrate the various methods of fitting measurement error models to paired observations. At best this approach is rather wasteful (it does not use all of the data in the most informative way), but at worst it can lead to results that are quite misleading. The use of estimated repeatability variances to calculate λ, which is then treated as if it were a known constant is an example of a sub-optimal, and possibly misleading, approach to the analysis of such data. The use of estimated repeatability variances to calculate the wrong λ (i.e. not accounting for inevitable equation error) is an obvious example of an invalid analysis.

If the collection of the data involves the use of replicate measurements from one or both of the measurements to be compared, then the subsequent statistical analyses should make full use of all of the data. Treating method comparison and repeatability as separate problems is inviting trouble. If nothing else, it is tempting us to ignore the pitfall of equation error. By all means explore the data in any way you like (including Bland–Altman plots and orthogonal regressions using means of the replicates), but the formal analysis and estimation of relative precision should be based on fitting the appropriate statistical model to the data as a whole. The advantage of the latter approach is that all of your assumptions have to be made explicit and open to scrutiny. I would have thought that this should be self-evident to all statisticians but, unfortunately, it is common to see statements claiming that the use of regression and measurement error modelling have no role to play in method comparison problems. Measurement error models for data arising from more informative designs will be the main topic for discussion in Chapter 4.

Returning to the analysis of pairs of observations, what is the recommended strategy for the analysis of the data? I will simply list what I think to be essential:

(a) Produce plots of the data. These should include simple X–Y scatter diagrams and the Bland–Altman plot. The Bland–Altman plot can be supplemented with estimated limits of agreement, and by the calculation of the correlation between the difference and mean. It might also be useful to look at box plots or other indicators of the marginal distributions of the observations.

(b) Produce simple summary statistics for X and Y, including means, variances and the covariance of X and Y. Look to see whether the covariance is larger than either of the two variances. If so, this is an indication that a simple model assuming equal scales of measurement (i.e. $\beta = 1$) will lead to an inadmissible (i.e. negative) error variance estimate. By all means estimate the product-moment correlation of X with Y, but do not expect it to necessarily be very informative. It might be useful to compare the product moment correlation with a measure of concordance such as the Krippendorff-Lin coefficient or the intra-class correlation.

(c) **IF** it is known *a priori* that $\beta = 1$, then proceed to use Grubbs' methods to estimate error variances and to test their equality. You may also wish to test whether $\alpha = 0$.

(d) **IF** the precision of the standard (i.e. σ_δ^2) is known, then carry out the appropriate regression analysis incorporating this assumption. Although it is superficially similar, be wary of assuming that you know the reliability ratio for the standard (i.e. κ_X). It may be appropriate to assume that σ_δ^2 is more or less the same in your present data set as it was in prior studies, but would it be credible to assume that the variance of the true scores (i.e. σ_τ^2) is also the same? It would be wise to check the sensitivity of the results to changes in the assumed value of σ_δ^2. The analysis might involve tests of equality of the error variances and/or whether $\beta = 1$ and/or whether $\alpha = 0$.

Note that, like Carroll & Ruppert (1996), I do not recommend the routine use of orthogonal regression. In effect this approach is based on the assumption that you know both σ_δ^2 and σ_ε^2 and hence their ratio, λ), but I hope that the reader realises by now that, given the almost certain presence of equation error, it is much safer to base regression analyses solely on assumptions concerning σ_δ^2. But whatever you do make sure that you have a large enough sample for the job at hand.

Method comparison 2: more informative designs

4.1 Introduction

We return to the basic measurement model for two methods, X and Y, that is:

$$X_i = \tau_i + \delta_i$$
$$Y_i = \alpha + \beta\tau_i + \varepsilon_i \tag{4.1}$$

We know that it is not identified as we only have a single measurement for each of the methods on each subject. We cannot estimate the parameters of the model, and much of Chapter 3 was taken up with methods of trying to get round this problem by the introduction of restrictive and possibly unrealistic assumptions. And we have also assumed that $\mathrm{Cov}(\delta_i, \varepsilon_i) = 0$. If this is not true then the situation we're in is even worse! So, how do we solve this problem? We collect more data using more informative designs. There are a range of related options including the use of instrumental variables, replication of the measurements using one or both of the methods, and using designs to compare more than two different methods. The resulting data are much richer than those discussed in Chapter 3, but, as well as solving the identification problem and enabling us to explore measurement process in more detail, they also pose quite a few analytical challenges. Measurement errors are more complicated than we naively might have first thought. We start with a discussion of instrumental variable methods.

4.2 The use of instrumental variables

If we can obtain measurements on a third variable Z, which need not itself be a measure of τ, but can be assumed to be correlated with it, and is also assumed to be uncorrelated with the measurement errors for X and Y, then we can solve at least part of the problem. Z is known as an instrumental variable. Let's describe the relationship between Z and τ by the following:

$$Z_i = \gamma + \lambda\tau_i + v_i \tag{4.2}$$

Note that we are not assuming that Z is a measure of τ but, in practice, this may often be the case (see following section). We assume that both $\mathrm{Cov}(v_i, \varepsilon_i)$ and $\mathrm{Cov}(v_i, \delta_i) = 0$ but, for the time being we drop the requirement that $\text{Cov}(\delta_i, \varepsilon_i) = 0$. Taken together, equations (4.1) and (4.2) imply the following expected values for X, Y and Z:

$$
\begin{aligned}
E(X_i) &= \mu \\
E(Y_i) &= \alpha + \beta\mu \quad \text{and} \\
E(Z_i) &= \gamma + \lambda\mu
\end{aligned}
\tag{4.3}
$$

The covariance matrix for X, Y and Z is

$$
\begin{bmatrix}
\sigma_\tau^2 + \sigma_\delta^2 & & \\
\beta\sigma_\tau^2 + \sigma_{\delta\varepsilon} & \beta^2\sigma_\tau^2 + \sigma_\varepsilon^2 & \\
\lambda\sigma_\tau^2 & \lambda\beta\sigma_\tau^2 & \lambda^2\sigma_\tau^2 + \sigma_\nu^2
\end{bmatrix}
\tag{4.4}
$$

where the notation is similar to that described in earlier chapters. Note that if we are aiming to estimate all of the parameters of this combined model ($\sigma_\tau^2, \sigma_\delta^2, \sigma_\varepsilon^2, \sigma_\nu^2, \sigma_{\delta\varepsilon}, \mu, \alpha, \gamma,$ β and λ) we cannot do this because the model is not identified. We are trying to estimate ten parameters using nine summary statistics. We can, however, estimate μ (from $\bar{X}$), β (from S_{YZ}/S_{XZ}) and α (from $\bar{Y} - \hat{\beta}\bar{X}$). The ratio S_{YZ}/S_{XZ} is referred to as the *instrumental variable (IV) estimate of* β. The asymptotic variance of $\hat{\beta}$ is given by

$$
\text{Var}(\hat{\beta}) = \frac{\sigma_u^2}{n\sigma_X^2\rho_{XZ}^2}
\tag{4.5}
$$

where n is the number of observations, σ_X^2 is the variance of X, σ_u^2 is the variance of the deviations $u_i = Y_i - \alpha - \beta X_i$ and ρ_{XZ}^2 is the square of the correlation between X and Z (see Fuller, 1987 or Wooldridge, 2000 for further details). Note that, keeping everything else constant, this variance decreases with increasing values of the correlation between X and the instrumental variable Z. Instrumental variable (two stage least squares) regression programs are available in many common general purpose statistical packages. Note, however, that these packages usually produce a significance test for β against H0: $\beta = 0$. In the present context we are much more likely to want to test against H0: $\beta = 1$. The components of the variance expression in (4.5) can easily be estimated. $n\sigma_X^2$ is simply the total sum of squares for X. ρ_{XZ}^2 can be obtained directly as the estimate of R^2 via ordinary least squares regression X on Z, and

$$
\hat{\sigma}_u^2 = \frac{1}{n-2}\sum_{i=1}^{n}\hat{u}_i^2
\tag{4.6}
$$

with the u_is being estimated using the IV estimates of α and β (i.e. $\hat{u}_i = Y_i - \hat{\alpha} - \hat{\beta}X_i$). Note that the variance estimate in (4.6) is not dependent upon any distributional assumptions.

Now, if we are prepared to return to the assumption that $\text{Cov}(\delta_i, \varepsilon_i) = 0$ then we can also estimate the three variances of direct interest as follows:

$$
\begin{aligned}
\hat{\sigma}_\tau^2 &= S_{XY}S_{XZ}/S_{YZ} \\
\hat{\sigma}_\delta^2 &= S_{XX} - S_{XY}S_{XZ}/S_{YZ} \\
\hat{\sigma}_\varepsilon^2 &= S_{YY} - S_{YZ}S_{XY}/S_{XZ}
\end{aligned}
\tag{4.7}
$$

The sampling variances of these estimates will be discussed in Section 4.4. We do not provide any of the estimators for the characteristics of the instrumental variable (e.g. λ and σ_v^2) because, for the time being at least, they are of no interest.

Before moving on, let's now consider an example. This has been described and analysed by Carter (1981). Carter's data are shown in Table 4.1(a). Carter is concerned with the measurement of the enzyme sucrase in intestinal tissues. The assay procedure first involves the homogenization of the tissue and the measurement of the specific activity in the homogenate. Then pellet fractions of the homogenates are prepared by centrifugation and the enzyme activities are measured in these pellet fractions. According to Carter, the pellet method is generally accepted as the standard method but it is time consuming and expensive to use. The homogenate method is a candidate as an alternative to the established pellet procedure. Carter's choice of an instrumental variable is a measurement of another enzyme, alkaline phosphatase, in the tissue homogenate. So, in line with our previous notation, X is a measurement of sucrase activity from the pellet, Y is a measure of sucrase from the homogenate and Z, the instrumental variable, is a measure of alkaline phosphatase, again from the homogenate. Following the recommendation of Fuller (1987), we first regress X on Z to check that Z is a contender as an instrument. The resulting R^2 is 0.464 ($R = 0.681$; $P < 0.001$). We then use two stage least squares to regress Y on X, using Z as an instrument. The IV estimate of β is 0.363 with a standard error of 0.056. The corresponding estimate of α is -4.67 (s.e. 9.58). Use of Equations (4.7) with the sample covariance matrix from Table 4.1(b) yields the following variance estimates

$$\hat{\sigma}_\tau^2 = \frac{S_{XY}S_{XZ}}{S_{YZ}}$$

$$= 2677.845 \times 4631.018/1681.174 = 7376.481$$

$$\hat{\sigma}_\delta^2 = S_{XX} - \frac{S_{XY}S_{XZ}}{S_{YZ}}$$

$$= 10031.120 - 7376.481 = 2654.639$$

$$\hat{\sigma}_\varepsilon^2 = S_{YY} - \frac{S_{YZ}S_{XY}}{S_{XZ}}$$

$$= 948.135 - (1681.174 \times 2677.485/4631.018)$$
$$= 948.135 - 971.124 = -22.989$$

Now we have a problem! The estimate of the variance of the errors for the homogenate ($\hat{\sigma}_\varepsilon^2$) is negative. This is not admissible. Barnett (1969) comments that a negative variance estimate such as this is embarrassing but not unprecedented. Having found a negative variance estimate, Carter (1981), Jaech (1985), Bolfarine and Galea-Rojas (1995), and others, infer that the true value of this variance is actually zero (the lower boundary for the permissible values) and proceed to estimate the other parameters of the model subject to this constraint. This may be mathematically the correct thing to do but in the context of the present application it is nonsense! There is no way that the homogenate method of measuring sucrase activity, for example, has zero measurement error variance. If an investigator were to conclude from analysis of data such as this that his measuring instrument is infallible (i.e. perfectly precise) it could be taken as a sign of madness. We need to step back from this finding and think what might have gone wrong. What is the likely cause of this embarrassment?

Table 4.1 Sucrose measurements in homogenate and pellet fractions, with alkaline phosphatase as an instrumental variable.

(a) Raw data

Subject	Homogenate	Pellet	Alkaline phosphatase
1	18.88	70.00	46.45
2	7.26	55.43	31.62
3	6.50	18.87	36.13
4	9.83	40.41	19.64
5	46.05	57.43	127.27
6	20.10	31.14	51.61
7	35.78	70.10	7.21
8	59.42	137.56	141.88
9	58.43	221.20	14.29
10	62.32	276.43	138.68
11	88.55	316.00	175.77
12	19.50	75.56	83.81
13	60.78	277.30	113.80
14	77.92	331.50	210.37
15	51.29	133.74	172.37
16	77.91	221.50	127.09
17	36.65	132.93	70.59
18	31.17	85.38	116.67
19	66.09	142.34	111.12
20	115.15	294.63	231.89
21	95.88	262.52	194.87
22	64.61	183.56	185.37
23	37.71	86.12	154.84
24	100.82	226.55	192.83

(b) Summary statistics

	Mean	Standard deviation
Homogenate	52.025	30.792
Pellet	156.175	100.155
Alkaline phospatase	114.836	67.884

Correlations

	Homogenate	Pellet	Alkaline phosphatase
Homogenate	1		
Pellet	0.868	1	
Alkaline phosphatase	0.804	0.681	1

Source: Carter (1981).

There is a possibility that the model (with its associated assumptions) may be the wrong one. Another likely possibility is that it is a sampling problem arising from the use of a pathetically small sample – this problem will be illustrated in detail below. But let's have a closer look at the data. Figure 4.1 provides a scatter plot matrix for the three sets of measurements. It looks as if subject number 9 may be an outlier. If we drop this possible outlier from the data and repeat the above analysis we obtain an estimate for σ_ε^2 of 24.297. This is still surprisingly low but at least it's not negative. If, instead of deleting the outlier, we fit a modified model with the intercept term, α, constrained to be zero (see Section 4.5 for details) then $\hat{\sigma}_\varepsilon^2 = 0.469$. Again this is unbelievably small but still not negative. You might be

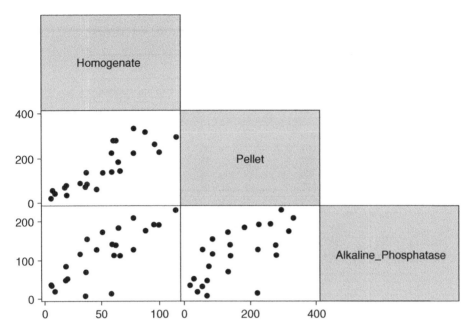

Figure 4.1 Scatterplot matrix for sucrose (from homogenate or pellet fractions) and alkaline phosphatase measurements.

convinced that the homogenate provides much more precise estimates than the standard pellet method, but the sample is so small that even this would not be a safe inference (see below).

Table 4.2 shows the results of a simple Monte Carlo simulation exercise to illustrate the problem of negative variance estimates. For each combination of parameter values I have produced 1000 simulated data sets and simply counted the number of negative variance estimates for the measurement errors in either X or Y. Sample sizes chosen were 50 or 200. In all simulations the following model held:

$$X = \tau + \delta$$
$$Y = \tau + \varepsilon \tag{4.8}$$
$$Z = \tau + \nu$$

with the variables τ, δ, ε and ν being independently normally distributed with zero mean and variances given below.

In estimating the relevant variances the correct model was assumed to be described by Equations (4.1) and (4.2) leading to the solutions given in Equation (4.7). In the first set of simulations (Table 4.2(a)), σ_τ^2 was equal to 9. The measurement error variances for both X and Y (i.e. σ_δ^2 and σ_ε^2, respectively) were fixed at 1. The variance of the random deviations for Z (i.e. σ_ν^2) varied from 1 to 16. When σ_ν^2 is low then Z is a relatively good instrument. As σ_ν^2 gets higher the value of the instrument decreases. This change can be reflected by the correlation between Z and τ or by the communality of Z (the first is the square root of the latter). The results in Table 4.2(a) speak for themselves. Negative variance estimates are very common and, unless we have a very good instrument, we are likely to need large sample sizes. But what if the measuring instruments were relatively much more precise (i.e. higher reliability) than those in Table 4.2(a)? Or, alternatively, what if they are much worse? Tables 4.2(b) and (c) illustrate the effect of changing σ_τ^2 to 99 or to 1, respectively,

Table 4.2 Monte Carlo simulations of the use of instrumental variables. (1000 simulations* for each combination of sample size and parameter values)

(a) $\sigma_\tau^2 = 9$

σ_ν^2	Corr(Z, τ)	Number of negative variance estimates	
		$N = 50$	$N = 200$
1	0.949	0	0
2	0.905	6	0
4	0.832	33	0
8	0.728	122	2
16	0.600	270	13

(b) $\sigma_\tau^2 = 99$

σ_ν^2	Corr(Z, τ)	Number of negative variance estimates	
		$N = 50$	$N = 200$
1	0.995	1	0
2	0.990	3	0
4	0.980	23	0
8	0.962	105	0
16	0.928	133	14

(c) $\sigma_\tau^2 = 1$

σ_ν^2	Corr(Z, τ)	Number of negative variance estimates	
		$N = 50$	$N = 200$
1	0.707	14	0
2	0.577	56	0
4	0.447	179	6
8	0.333	199	49
16	0.243	319	160

* Using EQS.

all other parameters taking the same values as in Table 4.2(a). The value of the instrumental variable (as measured by the number of negative variance estimates, at least) is not determined only by its correlation with the characteristic being measured. For a given correlation we are likely to do better if the variability in the characteristic being measured is increased. An illustration of the effect of the quality of the instrumental variable is further illustrated by Table 4.3. This provides summary statistics for estimates of ψ ($= \sigma_\delta^2 \beta^2/\sigma_\varepsilon^2$) and of ln ($\psi$) obtained from the simulations in the right hand column of Table 4.2(b). It is easy to see the dramatic effect of varying σ_ν^2 on the sampling distribution of these two estimates. Note also that, as might be expected, the distribution of the estimates of ψ is highly positively skewed, whereas that for its logarithm is roughly symmetrical. We will not go into any detail here, but it is quite straightforward, even without any distributional assumptions, to use jack-knife or bootstrap methods to produce standard errors or confidence intervals for the parameter estimates or of functions of two or more of these estimates.

Table 4.3 Summary statistics for relative precision ($\psi = \sigma_\delta^2\beta^2/\sigma_\varepsilon^2$) and $\ln(\psi)$ from data in the right hand column of Table 4.2(b).

(a) ψ

σ_ν^2	Obs	Mean	Std. Dev.	Min	Max
1	1000	1.016	0.254	0.425	2.507
2	1000	1.057	0.366	0.356	4.443
4	1000	1.103	0.566	0.221	7.025
8	1000	1.184	0.982	0.015	12.077
16	995*	1.822	5.982	0	163.912

(b) $\ln(\psi)$

σ_ν^2	Obs	Mean	Std. Dev.	Min	Max
1	1000	−0.013	0.245	−0.856	0.919
2	1000	0.001	0.329	−1.033	1.491
4	1000	−0.007	0.454	−1.509	1.949
8	1000	−0.063	0.694	−4.190	2.491
16	986**	−0.011	1.010	−5.918	5.099

* Not defined when $\hat{\sigma}_\varepsilon^2 \leqslant 0$.
** Not defined when either $\hat{\sigma}_\varepsilon^2 \leqslant 0$ or $\hat{\sigma}_\delta^2 \leqslant 0$.

4.3 Choice of instrumental variable

In the previous section we have seen an example of the use of instrumental variables in clinical chemistry. But, in general, what sort of measures might be used as instrumental variables? Clearly the answer is very context dependent, but I will describe a few possibilities. An obvious candidate is a third method of measurement of the same property as that measured by the first two. Another possibility is to repeat the measurements using either of the two original methods. For example, let's assume that we are interested in measuring the dietary nitrogen intake of each of a large sample of people taking part in an epidemiological survey. We wish to compare the measurement characteristics of a food frequency questionnaire with that of a one-week diary (both covering the same week of food intake). A contender for a third measurement method is some sort of biomarker (measurement of urinary nitrogen excretion, for example). But, instead, we might choose to wait a few weeks and repeat the one-week diary. The practical challenge is to ensure that (a) the instrumental variable is highly correlated with the true value of the property being measured by the first two methods, and (b) that it is not correlated with the measurement errors of the first two. A biomarker is likely to be a good candidate from the point of view of the second criterion (Wong *et al.*, 1999). Another possibility might be to use one of the original methods on a close family member (spouse or partner, for example, or siblings in the case of young children). Members of the same household are likely to have positively correlated dietary nitrogen intake (but, in practice, the correlation may not be high enough).

Now let's consider an example from clinical medicine. In estimating the efficacy of various forms of orally-administered drugs one is often interested in measuring the patient's degree of compliance with the clinician's recommendations. One might measure this by a simple self-report, for example, but also wish to compare the performance of this self-report with that of an electronically-monitored pill dispenser. One possibility for an instrumental variable is, again, some form of biomarker (levels of the drug or one of its metabolites in saliva, urine or blood, for example). But another is the patient's clinical response to treatment, a characteristic that should be correlated with intake of the drug. Finally, we might be able to measure something that might be a good predictor of compliance (we could ask the patient about his or her motivation to take the medication, or strength of belief in its effectiveness, for example). This situation is typical of many in the clinical, social and behavioural sciences. We may have two particular methods of measuring a specific characteristic or construct of interest, and then search for an instrumental variable that might predict the value of the target construct, or might be predicted by it. If we are lucky we may be able to find a third measure of the same construct but this may neither be possible nor necessary.

Fuller (1987, pp 56–8) illustrates instrumental variable methods from analyses of data on the reported magnitudes of Alaskan earthquakes for the period 1969 to 1978 (source: Meyers and von Hake, 1976). There are three measurements for each earthquake: the logarithm of the seismogram amplitude of 20-second surface waves, the logarithm of the seismogram amplitude of longitudinal body waves, and the logarithm of the maximum seismogram trace amplitude at short distance. In this example the choice of which of the three measurements should be regarded as the two for comparison, and which should be the instrumental variable is rather arbitrary, and the approach to the analysis in this case is better described in the following section on the comparison of the performance of three or more measuring instruments. Fuller (1987, pp 63–8) also discusses another example from a 1978 study of the U.S. Department of Agriculture in which the area of land under specific crops was assessed by digitized aerial photography, satellite imagery, and by interviewing the farmer. If the main aim of the statistical analysis were to compare the measurement characteristics of aerial photography and satellite images, then the farmer's report fulfills the role of the instrumental variable.

4.4 Three methods: no replication

Consider a method comparison study in which we have assessed each of n randomly selected subjects or specimens using three different measuring instruments or assays. Let the standard assay be X as before, and the two comparison assays be Y and Z. The measurement model that we adopt is of the following form. For the ith subject or specimen:

$$X_i = \tau_i + \delta_i$$
$$Y_i = \alpha_y + \beta_y \tau_i + \varepsilon_i \qquad (4.9)$$
$$Z_i = \alpha_z + \beta_z \tau_i + \upsilon_i$$

where the measurement errors (δ, ε and υ) are all mutually independent, both within and between subjects. The measurement errors are also independent of the true values (τ). If $E(\tau_i) = \mu$ then:

$$E(X_i) = \mu$$
$$E(Y_i) = \alpha_y + \beta_y \mu \qquad (4.10)$$
$$E(Z_i) = \alpha_z + \beta_z \mu$$

This model also implies the following expected covariance matrix:

$$\begin{bmatrix} \sigma_\tau^2 + \sigma_\delta^2 & & \\ \beta_y\sigma_\tau^2 & \beta_y^2\sigma_\tau^2 + \sigma_\varepsilon^2 & \\ \beta_z\sigma_\tau^2 & \beta_y\beta_z\sigma_\tau^2 & \beta_z^2\sigma_\tau^2 + \sigma_\nu^2 \end{bmatrix} \tag{4.11}$$

This is mathematically identical to that provided in Equations (4.1) and (4.2), the only difference being the replacement of γ and λ in (4.2) by α_z and β_z, respectively, to explicitly recognize that either Z can be considered as the instrumental variable for the relationship between Y and X, or Y can be considered as the instrument for the relationship between Z and X. In fact, in an entirely symmetrical way, X can be considered as the instrument for the relationship between Y and Z. Unlike the earlier situation, we now interpret σ_ν^2 as a variance of measurement errors.

Now, given a set of data on X, Y and Z we can estimate the above parameters by the method of moments, irrespective of any distributional assumptions. As before, we simply calculate the sample means of X, Y and Z and equate them with their expected values. We then calculate the sample variances and covariances and again equate these with their expected values. The solutions obtained are given in Table 4.4(a). Again, without making any distributional assumptions we can use bootstrap sampling to obtain standard errors or confidence intervals for the parameter estimates. Alternatively, we might assume that the τ, δ, ε and ν are jointly multivariate normal, and proceed to use maximum likelihood methods to estimate the required standard errors (see following section).

We can also estimate the precision of each of the three instruments and also the precision of either Y or Z relative to that of, say, the standard, X. As before, their standard errors can be estimated using bootstrap sampling. Consider Table 4.5. This contains the raw data and summary statistics from three measures of serum cholesterol from a paper by Morton & Dobson (1989). Although the sample size ($n = 24$) is very small, we use these data to illustrate the methodology. Two-way scatterplots for these data are shown in Figure 4.2. Use of the estimators provided in Table 4.4(a) yields the estimates given in the second column of Table 4.6. The sampling variability of these estimates can be seen by reference to the columns to the right of these estimates. These are based on 100 bootstrap samples only, but as in Section 4.2, do give a good idea of the problem we are facing. Again we emphasize that we need large samples.

4.5 Model-fitting and inference

In the above section we have fitted a model (i.e. estimated the model's parameters) by simply equating the observed and expected moments (means, variances and covariances) then solving the resulting simultaneous equations to obtain the estimates of the parameters. Typically, however, we will find parameter estimates by using iterative numerical methods involving maximizing a predefined goodness-of-fit criterion. In general, we will represent an observed mean by the vector $\boldsymbol{m}$ and it's predicted value by the vector $\hat{\mu}$. Similarly, we will represent an observed covariance matrix by S and the covariance matrix predicted by the parameter estimates by $\hat{\boldsymbol{\Sigma}}$. The most commonly used fitting criterion involves minimizing

$$F_{ML} = \frac{1}{2}\left[\log|\hat{\boldsymbol{\Sigma}}| + tr(\hat{\boldsymbol{\Sigma}}^{-1}\mathbf{T}) - \log|S| - q\right] \tag{4.12}$$

Table 4.4 Estimation of the parameters of the model for three measures: X, Y and Z (Barnett, 1969).

(a) Moments estimators

$$\hat{\mu} = \overline{X}$$
$$\hat{\beta}_Y = S_{YZ} / S_{XZ}$$
$$\hat{\beta}_Z = S_{YZ} / S_{XY}$$
$$\hat{\alpha}_Y = \overline{Y} - \hat{\beta}_Y \hat{\mu}$$
$$\hat{\alpha}_Z = \overline{Z} - \hat{\beta}_Z \hat{\mu}$$
$$\hat{\sigma}_\tau^2 = S_{XY} S_{XZ} / S_{YZ}$$
$$\hat{\sigma}_\delta^2 = S_{XX} - \hat{\sigma}_\tau^2$$
$$\hat{\sigma}_\varepsilon^2 = S_{YY} - \hat{\beta}_Y^2 \hat{\sigma}_\tau^2$$
$$\hat{\sigma}_\nu^2 = S_{ZZ} - \hat{\beta}_Z^2 \hat{\sigma}_\tau^2$$

(b) Approximate variances of the parameter estimates

$$\mathrm{Var}(\hat{\mu}) = \left(\sigma_\tau^2 + \sigma_\delta^2\right)/n$$
$$\mathrm{Var}(\hat{\sigma}_\tau^2) = \left[2\sigma_\tau^2\left(\sigma_\tau^2 + 2\sigma_\delta^2\right) + \Delta\right]/n$$
$$\mathrm{Var}(\hat{\beta}_Y) = \frac{1}{n\sigma_\tau^2}\left(\sigma_\varepsilon^2 + \sigma_\delta^2\beta_Y^2\right)\left(1 + \sigma_\varepsilon^2 / \beta_Y^2\sigma_\tau^2\right)$$
$$\mathrm{Var}(\hat{\beta}_Z) = \frac{1}{n\sigma_\tau^2}\left(\sigma_\nu^2 + \sigma_\delta^2\beta_Z^2\right)\left(1 + \sigma_\nu^2 / \beta_Z^2\sigma_\tau^2\right)$$
$$\mathrm{Var}(\hat{\alpha}_Y) = \left(\beta_Y^2\sigma_\delta^2 + \sigma_\varepsilon^2\right)/n + \mu^2\mathrm{Var}(\hat{\beta}_Y)$$
$$\mathrm{Var}(\hat{\alpha}_Z) = \left(\beta_Z^2\sigma_\delta^2 + \sigma_\nu^2\right)/n + \mu^2\mathrm{Var}(\hat{\beta}_Z)$$
$$\mathrm{Var}(\hat{\sigma}_\delta^2) = \left(2\sigma_\delta^4 + \Delta\right)/n$$
$$\mathrm{Var}(\hat{\sigma}_\varepsilon^2) = \left(2\sigma_\varepsilon^4 + \beta_Y^4\Delta\right)/n$$
$$\mathrm{Var}(\hat{\sigma}_\nu^2) = \left(2\sigma_\nu^4 + \beta_Z^4\Delta\right)/n$$

where
$$\Delta = \sigma_\tau^2\sigma_\varepsilon^2/\beta_Y^2 + \sigma_\tau^2\sigma_\nu^2/\beta_Z^2 + \sigma_\varepsilon^2\sigma_\nu^2/\beta_Y^2\beta_Z^2$$

where the matrix $\boldsymbol{T}$ is given by

$$\boldsymbol{T} = \boldsymbol{S} + (\boldsymbol{m} - \hat{\mu})(\boldsymbol{m} - \hat{\mu})'$$

and q is the number of latent variables or factors (see Jöreskog and Sörbom, 1979; Sörbom, 1982; Browne and Arminger, 1995). The symbol 'tr' is the trace operator indicating the sum of the diagonal elements of a matrix. F_{ML} produces estimates that are maximum likelihood, provided that the data are multivariate normal. Strictly speaking, for the variance estimators to be ML we should estimate Σ by dividing the sums of squares or sums of products by the sample size, n, rather than the more usual $n - 1$. For reasonably-sized samples, however, the difference is trivial. Other fitting criteria are discussed in Bollen (1989).

Table 4.5 Measurements of serum cholesterol.

(a) Raw data

Subject	Method 1	Method 2	Method 3
1	6.10	6.81	6.50
2	2.97	2.59	3.50
3	4.90	4.81	5.40
4	5.13	5.12	4.90
5	5.07	5.21	5.20
6	5.66	5.62	5.60
7	5.57	5.48	5.80
8	4.07	3.96	4.30
9	4.95	4.78	5.00
10	4.65	4.70	4.90
11	6.09	6.14	6.60
12	8.41	8.00	7.90
13	5.39	5.42	5.50
14	5.74	6.10	5.80
15	4.26	4.00	3.90
16	5.43	5.23	5.70
17	5.04	4.95	5.40
18	3.73	3.85	4.50
19	4.75	4.54	5.10
20	4.98	5.17	5.50
21	5.95	5.98	6.20
22	6.45	6.61	6.70
23	5.25	5.44	5.70
24	5.59	5.93	6.10

(b) Summary statistics

	Mean	Standard deviation
Method 1	5.2554	1.0403
Method 2	5.2683	1.1073
Method 3	5.4875	0.9502

Correlation/Covariance matrix*

Method 1	Method 2	Method 3
1.0821	0.9752	0.9583
1.1234	1.2263	0.9658
0.9472	1.0162	0.9029

* Covariances below diagonal; variances on the diagonal; correlations above
Source: Morton, A.P. and Dobson, A.J. Assessing agreement. *MJA 1989*, 150: 384–7. © Copyright 1989. *The Medical Journal of Australia* – reproduced with permission.

It clearly makes sense to use the optimal value of the fitting criterion, such as the minimum value of F_{ML}, as a measure of the goodness of fit of the model: the lower the value the better the fit. In fact, assuming multivariate normality, it is possible to show that F_{ML} has (for sufficiently large sample sizes) a chi-squared distribution under the null hypothesis that the covariance matrix is of the form predicted by the model. The degrees of freedom for the chi-squared distribution (d.f.) is given by

$$\text{d.f.} = p + p(p + 1)/2 - t \tag{4.13}$$

Table 4.6 Parameter estimates from the cholesterol data.

Parameter	Estimate	Bootstap results (100 replications)				s.e.	
		Mean	s.d.	5th%	95th%	ML	Robust
α_y	−0.370	−0.470	0.447	−1.268	0.302	0.277	**
α_z	0.734	0.688	0.456	−0.434	1.295	0.301	**
μ	5.255	5.251	0.195	4.951	5.586	0.217	**
β_y	1.073	1.095	0.089	0.946	1.258	0.052	0.082
β_z	0.905	0.915	0.086	0.796	1.124	0.056	0.065
σ^2	1.047	1.025	0.406	0.362	1.739	0.319	0.442
σ_δ^2	0.035	0.014	0.028	0.006	0.049	0.016	0.016
σ_ε^2	0.021	0.013	0.019	0.000*	0.045	0.016	0.013
σ_ν^2	0.046	0.017	0.041	0.014	0.069	0.017	0.018
ψ_y	1.919	5.168	9.709	0.285	33.85		
ψ_z	0.623	0.709	0.700	0.119	1.797		

* Negative variance estimate – variance arbitrarily set at 0.001 when calculating ψ.
** Not available in EQS.

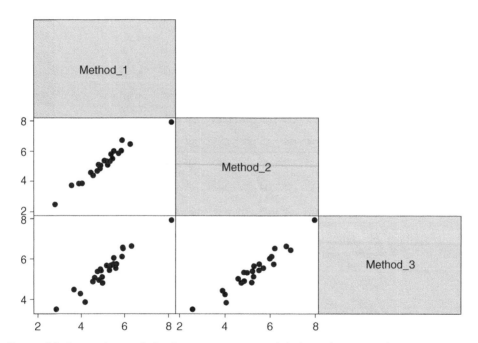

Figure 4.2 Scatterplot matrix for three measurements of cholesterol concentration.

where p is the number of measured variables and t the number of estimated (free) parameter values.

If we have measurements that are clearly not normally distributed then we have to be cautious about the interpretation of goodness of fit (chi-squared) statistics and the standard errors of the parameter estimates. One possibility is to transform the data to normality (or at least symmetry) but this is not always possible, particularly with data such as severity measures, which often have a modal value at zero. Another possibility is to estimate the parameters of the model using the inappropriate maximum likelihood criterion, but then to obtain estimates of the standard errors by a method robust to this incorrect model specification. Jack-knife and bootstrap (as in the previous section) are two possibilities. A more easily implemented approach is to use a 'sandwich' estimator of the covariance parameter estimates such as that of Huber (1967), White (1982), Browne and Shapiro (1988) or Satorra and Bentler (1990). When reporting robust standard errors we will use the method of Satorra and Bentler.

Staying, for the time being, with models involving only three measurements, and assuming multivariate normality, Barnett (1969) derived ML estimators for each of the parameters of the model described by Equation (4.9). In this particular case (i.e. three measurement instruments) these are identical to those already derived from equating observed and expected moments (i.e. means and covariance matrices – see Table 4.4(a)). Barnett (1969) also derived approximate variances for these parameter estimates. These are given in Table 4.4(b). When we have more than three measurements, or when we have a model for a more complex data set, then we use standard large sample results to estimate the variances (and covariances) of the estimates of the model parameter. Software is widely available for fitting covariance structure models such as those described here, and common examples are listed in Appendices 2, 3 and 4. Our illustrative examples in this chapter will involve the use of the software packages *EQS* (Bentler, 1995), *Mplus* (Muthén and Muthén, 1998-2001) or the *Stata* command *gllamm* (Rabe-Hesketh *et al.*, 2002, 2004).

Using ML estimation in EQS, the standard errors for the parameters of the measurement model from the cholesterol data are shown on the right hand side of Table 4.6 These are accompanied by the corresponding robust estimates (the column on the far right of the table).

4.6 Testing hypotheses (constraints)

Now let's return to have a more in-depth look at the string data first described in Chapter 1 (Table 1.2; see also Section 2.5). For the time being we will ignore the rule and start by fitting a model such as that given in Equation (4.9) to the guesses of Graham, Brian and Andrew. That is, we fit the following:

$$G_i = \tau_i + \delta_i$$
$$B_i = \alpha_B + \beta_B \tau_i + \varepsilon_i \qquad (4.14)$$
$$A_i = \alpha_A + \beta_A \tau_i + \upsilon_i$$

The parameter estimates provided by *EQS* are given in Table 4.7(a). The model is a perfect fit. Although we cannot test any of the assumptions of the model from these data, we can introduce constraints on the model's parameters to ask questions concerning the relative performance of the three measuring 'instruments'. Perhaps the obvious place to start is by constraining all of the intercept terms to be zero (none of the estimates of the αs is statistically significantly different from zero, and in any case one would expect that the observers would see when the lengths of the pieces of string were shrinking towards nothing). The results of fitting the revised model are shown in Table 4.7(b). The chi-square goodness-of-fit statistic provides a test of the constraint on the intercept terms (4.466 with 2 d.f.). The revised model fits pretty well. The next obvious hypothesis to test is that the slopes of the three lines are the same (that is, equal to 1) in addition to the intercepts being zero. Here, however we obtain a chi-square statistic of 18.505 with 4 d.f. The model does not fit (P < 0.001). The difference between this chi-square and the previous one is

Table 4.7 Results of fitting measurement models to the guesses of the lengths of the 15 pieces of string by Graham, Brian and Andrew.

(a) The unconstrained measurement model (Equation 4.14)

Parameter	Estimate (s.e.)	Parameter	Estimate (s.e.)
α_G	0*	β_G	1*
α_B	−0.367 (0.337)	β_B	1.105 (0.091)
α_A	−0.532 (0.254)	β_A	1.252 (0.069)
σ^2_G	0.069 (0.038)		
σ^2_B	0.150 (0.066)		
σ^2_A	0.019 (0.044)		
μ_T	3.433 (0.389)	σ^2_T	2.052 (0.802)
Reliablities:			
ρ_G	0.967		
ρ_B	0.943		
ρ_A	0.994		

(b) After constraining all intercepts to be zero

Parameter	Estimate (s.e.)	Parameter	Estimate (s.e.)
α_G	0*	β_G	1*
α_B	0*	β_B	1.015 (0.034)
α_A	0*	β_A	1.121 (0.028)
σ^2_G	0.081 (0.047)		
σ^2_B	0.140 (0.065)		
σ^2_A	0.046 (0.048)		
μ_T	3.382 (0.419)	σ^2_T	2.393 (0.916)
Reliablities:			
ρ_G	0.967		
ρ_B	0.946		
ρ_A	0.985		

* Constraints.

(c) Fitting revised model (Equation 4.15) after constraining all intercepts to be zero

Parameter	Estimate (s.e.)	Parameter	Estimate (s.e.)
α_G	0*	β_G	1*
α_B	0*	β_B	1.015 (0.034)
α_A	0*	β_A	1.121 (0.028)
σ^2_G	0.081 (0.047)		
σ^2_B	0.136 (0.064)		
σ^2_A	0.036 (0.039)		
μ_T	3.382 (0.419)	σ^2_T	2.388 (0.914)
Reliablities:			
ρ_G	0.967		
ρ_B	0.946		
ρ_A	0.985		

(continued)

Table 4.7 (*continued*).

(d) Fitting revised model (Equation 4.14) after constraining all intercepts to be zero and measurement error variances to be equal

Parameter	Estimate (s.e.)	Parameter	Estimate (s.e.)
α_G	0*	β_G	1*
α_B	0*	β_B	1.018 (0.030)
α_A	0*	β_A	1.119 (0.033)
σ_G^2	0.086 (0.023)		
σ_B^2	0.086 (0.023)		
σ_A^2	0.086 (0.023)		
μ_T	3.389 (0.415)	σ_T^2	2.338 (0.898)
Reliablities:			
ρ_G	0.965		
ρ_B	0.965		
ρ_A	0.965		

* Constraints.

$18.505 - 4.466 = 14.038$ with 2 degrees of freedom. We conclude that the slopes are not equal.

We could now test whether the measurement error variances were equal, but although this would be very straightforward to carry out, we need to ask whether this would be a sensible thing to do. Because the slopes differ (that is, the scales of measurement for the three observers differ) the error variances are not strictly comparable. We need instead a test for equal precisions (that is whether the β^2/σ^2 ratios are the same or, equivalently, are the reliabilities). We start be reformulating our model as follows:

$$G_i = \tau_i + \delta_i$$
$$B_i = \alpha_B + \beta_B(\tau_i + \varepsilon_i) \qquad (4.15)$$
$$A_i = \alpha_A + \beta_A(\tau_i + \upsilon_i)$$

In this model we apply the same scaling factor (β coefficient) to both the true values and the random measurement errors and these βs now drop out of the expressions for both the precision and the reliability of each of the observers. Fitting this model with the intercepts all constrained to zero gives the results in Table 4.7(c). The job set up in *EQS* is given in Appendix 2(a). Note that we get the same chi-square (4.466 with 2 d.f.) as that for Table 4.7(b). The only changes are in the estimates of the four variances (that of the true score and those for the three measurement errors). Now that we constrain the error variances to be equal we are carrying out a test of the equality of the three instruments' precisions (or reliabilities). The results are given in Table 4.7(d). The resulting chi-square is 6.419 with 4 d.f. (P = 0.170). The data are consistent with the three observers having the same precision (but as noted earlier we must bear in mind the very low power for this test based on only observations on 15 pieces of string).

4.7 Four or more methods: no replication

Ideally, a method comparison should include more than three instruments (methods). As we have seen above, the unconstrained measurement model involving only three methods provides a perfect fit to the data, but there is no way of checking any of the assumptions needed for the validity of the estimates (that the measurement errors for the different methods are uncorrelated, for example). With four or more we have degrees of freedom to check goodness-of-fit. We will illustrate this by going back to the string data once again but this time include measurements provided by the Rule (R) in the model. We fit the following in *EQS*:

$$
\begin{aligned}
R_i &= \tau_i + \omega_i \\
G_i &= \alpha_G + \beta_G \tau_i + \delta_i \\
B_i &= \alpha_B + \beta_B \tau_i + \varepsilon_i \\
A_i &= \alpha_A + \beta_A \tau_i + \upsilon_i
\end{aligned}
\tag{4.16}
$$

in which the notation should be obvious from the context. The results are shown in Table 4.8(a). The resulting goodness-of-fit is indicated by a chi-square statistic of 2.348 with 2 d.f. The model is a very good fit (but remember that the sample size is very small). Note that the measurement error variance for the Rule is much smaller than those for the three subjective measurements. Bearing in mind the earlier warning that no measurement instrument is perfectly precise, we could assume that for all practical purposes in the present exercise there are no errors in the measurements provided by the Rule (and, in fact, once

Table 4.8 Analysis of the full string data set (i.e. 4 methods).

(a) After fitting Equation (4.16) using *EQS*

Parameter	Estimate (s.e.)	Parameter	Estimate (s.e.)
α_R	0*	β_R	1*
α_G	0.312 (0.183)	β_G	0.727 (0.039)
α_B	−0.062 (0.209)	β_B	0.813 (0.044)
α_A	−0.125 (0.163)	β_A	0.906 (0.035)
σ_R^2	0.009 (0.020)		
σ_G^2	0.076 (0.031)		
σ_B^2	0.100 (0.041)		
σ_A^2	0.056 (0.027)		
μ_T	4.293 (0.526)	σ_T^2	3.869 (1.466)**

(b) After fitting Equation (4.16) using *gllamm*

Parameter	Estimate (s.e.)	Parameter	Estimate (s.e.)
α_R	0*	β_R	1*
α_G	0.313 (0.177)	β_G	0.727 (0.038)
α_B	−0.062 (0.199)	β_B	0.813 (0.042)
α_A	−0.124 (0.157)	β_A	0.906 (0.034)
$\log(\sigma_R)$	−2.438 (1.342)		
$\log(\sigma_G)$	−1.316 (0.213)		
$\log(\sigma_B)$	−1.192 (0.192)		
$\log(\sigma_A)$	−1.467 (0.261)		
μ_T	4.293 (0.491)	σ_T^2	3.611 (1.321)**

*Constraint.
**These differ because the EQS run is based on summary statistics using $(n-1)$ as the denominator for estimation of sample variances and covariances; the *gllamm* run, based on the raw data is using ML (i.e. equivalent to dividing by $n - 3.869 \times 14/15 = 3.611$).

we start introducing constraints into this model we often find that this error variance is estimated to be zero). If we constrain the Rule's measurement errors to have zero variance, the resulting chi-square is 2.487 with 3 d.f. (the change is clearly negligable!). Note that in fitting this model we have simply undertaken three simultaneous linear regressions (Graham against the Rule, Brian against the Rule and Andrew Against the Rule).

Instead of using software for the analysis of covariance structures (*EQS*, for example) we can fit these measurement models as examples of generalized linear latent and mixed models (GLLAMMs) using the *gllamm* command in *Stata*. We will not dwell on the technical details of these approaches here but simply illustrate how it is done and provide the results. Approaching the problem through the use of *gllamm* will allow us some more flexibility in the section on heteroscedastic errors to follow. Table 4.8(b) provides the results of using *gllamm* to fit Equation (4.16) to the string data. Note that the variability of the measurement errors is parameterized in terms of the logarithm of their standard deviation. Otherwise, the results are comparable to those produced using *EQS* or other covariance structure software programs. The *gllamm* command for this analysis is given in Appendix 3(a).

4.8 Heteroscedastic error variances

In fitting the measurement models to the string data in Table 1.2 we have assumed that the variance of the measurement errors is constant for a given observer. That is, we have assumed that it does not change with the true length of the piece of string being observed. But it is fairly clear *a priori* that the variability of repeated guesses will be higher for the longer pieces of string than for those that are very short. However, we have very little data on which to check whether this might be true. But consider Table 4.9. This contains data from measurements of the testicular volume of 42 adolescents using five different techniques. The data were collected as part of a study by Chipkevitch *et al.* (1996) and are produced in Table 1 of Galea-Rojas *et al.* (2002). The data have been re-ordered according to the volume given using the standard method (*US*), which is based on the use of ultrasound (the subject numbers are as in Galea-Rojas *et al.*). The other five, in order, are a graphical method (*G*), a dimensional method (*D*), one based on the use of a Prader orchidometer (*P*) and another based on the use of a ring orchidometer (*R*). Summary statistics are presented in Table 4.10(a). A scatter plot matrix is given in Figure 4.3. Note that the measurements appear to be positively skewed and, more interestingly for the present discussion, the scatter appears to increase with the recorded volumes. Galea-Rojas *et al.* (2002) carried out a variance-stabilizing transformation (a cube root) prior to fitting measurement models to their data. Summary statistics for the cube roots of the five volume assessments are given in Table 4.10(b) and a scatter plot matrix is shown in Figure 4.4.

We start here by fitting models to summary statistics obtained from the raw data (despite the obvious breakdown of the assumption of constant error variance within each method). The basic model has the form:

$$US_i = \tau_i + \omega_i$$
$$G_i = \alpha_G + \beta_G \tau_i + \delta_i$$
$$D_i = \alpha_D + \beta_D \tau_i + \varepsilon_i$$
$$P_i = \alpha_P + \beta_P \tau_i + \upsilon_i \qquad (4.17)$$
$$R_i = \alpha_R + \beta_R \tau_i + \xi_i$$

Table 4.9 Testicular volume data (in ml).

Subject	US	G	D	P	R
4	2.6	3.5	3.1	4.0	4.0
20	2.7	3.5	4.1	2.5	6.0
14	3.0	2.0	2.0	3.0	4.0
18	4.1	3.5	4.4	4.0	6.0
23	4.5	3.5	3.9	6.0	7.0
1	5.0	7.5	5.9	8.0	9.0
30	5.3	5.0	5.9	8.0	10.0
28	5.4	5.0	6.1	8.0	8.0
24	5.6	5.0	4.5	4.5	6.0
2	5.7	5.0	4.8	6.0	10.0
5	5.7	5.0	5.0	6.0	7.0
6	6.1	5.0	4.4	7.0	8.0
7	6.2	5.0	6.0	8.0	9.0
29	6.7	7.5	7.2	10.0	8.0
3	7.4	5.0	6.8	9.0	12.0
27	8.5	7.5	8.8	12.0	12.0
42	8.9	10.0	8.1	12.0	12.0
9	9.1	7.5	7.9	10.0	11.0
35	9.1	7.5	10.8	12.0	12.0
26	9.2	10.0	11.3	12.0	13.5
37	9.3	10.0	8.4	10.0	10.0
34	9.4	10.0	10.3	12.0	13.5
12	9.6	7.5	8.2	10.0	11.0
40	9.7	10.0	9.7	11.0	12.0
17	10.0	7.5	7.9	12.0	12.0
21	10.2	10.0	11.1	12.0	13.5
8	10.4	10.0	8.8	10.0	10.0
25	11.0	7.5	9.7	9.0	11.0
39	11.5	10.0	10.6	15.0	13.5
19	12.7	10.0	11.4	12.0	12.0
41	13.7	12.5	11.6	17.5	15.0
33	13.9	12.5	12.2	15.0	17.5
36	14.1	15.0	13.0	13.5	15.0
10	14.8	10.0	13.0	12.0	15.0
13	15.7	15.0	19.8	20.0	20.0
11	16.4	12.5	10.3	17.5	17.5
15	16.4	15.0	17.3	20.0	20.0
22	16.5	10.0	15.3	15.0	15.0
16	17.6	15.0	17.3	20.0	22.5
32	18.8	15.0	16.3	20.0	25.0
31	20.0	20.0	16.3	25.0	22.5
38	20.9	20.0	22.1	25.0	25.0

Source: Galea-Rojas *et al.* (2002).

where the notation should be clear from the context. The results of fitting this model via maximum likelihood (assuming multivariate normality) in *EQS* are shown in Table 4.11(a). The resulting chi-square goodness-of-fit statistic is 5.070 with 5 d.f. The error variances all look fairly similar, and the slope estimates are all fairly close to 1. The intercepts are all close to 0 except that for the ring orchidometer (*R*). If we refit the model, constraining all five slopes to be equal to 1, we obtain a chi-square statistic of 29.651 with 9 d.f. (the change being 24.581 with 4 d.f.: P < 0.001). So, although the slopes look to be fairly close to 1 there are statistic-ally significant differences. If, instead, we refit the original model with the intercept terms all constrained to be 0, we obtain a chi-square statistic of 13.191 with 9 d.f. (a change of 8.121 with 4 d.f.: P > 0.05). The differences in intercept estimates may be due to chance variation.

Table 4.10 Summary statistics for testicular volumes ($N = 42$).

(a) Raw data

	US	G	D	P	R
Mean	10.0810	9.0119	9.5619	11.5595	12.4524
Standard deviation	4.9156	4.3777	4.7182	5.5592	5.3062
Covariances/Correlations*					
US	–	0.9368	0.9406	0.9476	0.9471
G	20.1588	–	0.9262	0.9531	0.9269
D	21.8154	19.1297	–	0.9348	0.9383
P	25.8963	23.1944	24.5182	–	0.9592
R	24.7039	21.5311	23.4908	28.2956	–

* Covariances below the diagonal, correlations above.

(b) After taking the cube root of the volumes

	US	G	D	P	R
Mean	2.1005	2.0260	2.0649	2.2007	2.2722
Standard deviation	0.3650	0.3426	0.3540	0.3760	0.3288
Covariances/Correlations*					
US	–	0.9394	0.9499	0.9484	0.9454
G	0.1775	–	0.9439	0.9456	0.9257
D	0.1227	0.1145	–	0.9406	0.9468
P	0.1302	0.1218	0.1252	–	0.9550
R	0.1135	0.1043	0.1102	0.1181	–

* Covariances below the diagonal, correlations above.

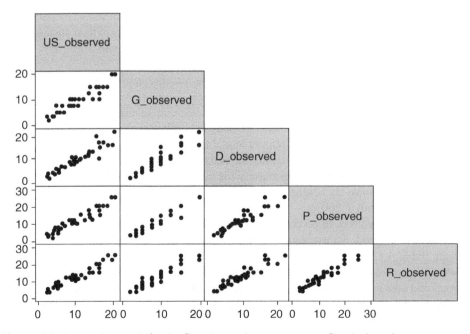

Figure 4.3 Scatterplot matrix for the five observed measurements of testicular volume.

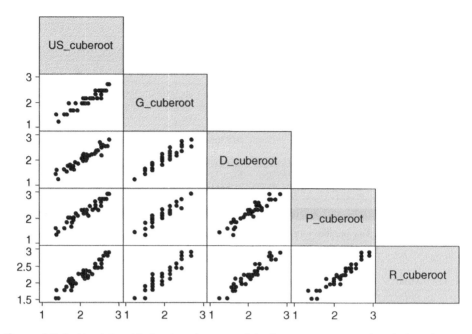

Figure 4.4 Scatterplot matrix for the cube-roots of the five measurements of testicular volume.

Now let's look at a similar analysis of the data following a cube root transformation of the measured volumes. The parameter estimates are shown in Table 4.11(b). Qualitatively, they convey a similar message to that in Table 4.11(a). The goodness-of-fit chi-square is 4.637 with 5 d.f. Constraining the slopes all to be 1 gives a chi-square of 12.969 with 9 d.f. (the change being 8.332 with 4 d.f.: P > 0.05). This result does not appear to be consistent with our earlier test based on the raw data. Constraining the intercepts to be 0, instead, provides a chi-square of 20.008 with 9 d.f. (the change being 15.361 with 4 d.f.: P < 0.01). The intercepts differ!

To summarize the present position, fitting the linear measurement models to the raw data (possibly the most sensible model but with unrealistic assumptions concerning the distribution of the errors) and to the transformed data (a less convincing model but with more realistic assumptions concerning the errors) leads to conflicting results. Where do we go from here? We try fitting our measurement model to the raw data, as before, but this time we make sensible assumptions concerning the behaviour of the errors. We let the variance of the errors within each method increase with increasing volume. In the psychometric literature this is known as heteroscedastic factor analysis (Meijer & Mooijaart, 1996; Lewin-Koh & Amemiya, 2003).

Unfortunately, we know of no commercially available software that will enable us to do this with ease. Instead, we use a simple approximation and fit the modified model using *gllamm*. Table 4.12(a) gives the estimates for fitting Equation (4.17) to the raw data in *gllamm*. We now calculate the mean of the five volume measurements for each of the 42 subjects and take its logarithm. We then centre the values of the logged mean (i.e. so that it now has a mean of zero). Finally, we re-run the estimation in *gllamm* after also specifying that the logarithm of the standard deviation measurement errors (i.e. the form in which *gllamm* estimates error variability) is linearly related (with slope γ) to the centred logarithm

Table 4.11 Results of fitting basic measurement model to testicle volumes (in *EQS*).

(a) Using raw data

Parameter	Estimate (s.e.)	Parameter	Estimate (s.e.)
α_{US}	0*	β_{US}	1*
α_G	0.103 (0.574)	β_G	0.884 (0.052)
α_D	−0.008 (0.632)	β_D	0.949 (0.057)
α_P	0.045 (0.633)	β_P	1.142 (0.057)
α_R	1.540 (0.643)	β_R	1.082 (0.058)
σ^2_{US}	1.372 (0.379)		
σ^2_G	1.364 (0.357)		
σ^2_D	1.722 (0.444)		
σ^2_P	1.168 (0.373)		
σ^2_R	1.451 (0.413)		
μ_T	10.081 (0.768)	σ^2_T	22.791 (5.333)

(b) Using cube-roots

Parameter	Estimate (s.e.)	Parameter	Estimate (s.e.)
α_{US}	0*	β_{US}	1*
α_G	0.073 (0.112)	β_G	0.930 (0.053)
α_D	0.031 (0.109)	β_D	0.968 (0.051)
α_P	0.033 (0.112)	β_P	1.032 (0.053)
α_R	0.383 (0.102)	β_R	0.899 (0.048)
σ^2_{US}	0.007 (0.002)		
σ^2_G	0.008 (0.002)		
σ^2_D	0.007 (0.002)		
σ^2_P	0.007 (0.002)		
σ^2_R	0.006 (0.002)		
μ_T	2.100 (0.057)	σ^2_T	0.126 (0.029)

of the mean of each subject's five measurements. The latter is an approximation for the subject's unknown true value. The estimated intercepts from this part of the model give us information on the variability of the errors as we pass from one method to another, and the common slope tells us how this variability increases with volume. In terms of the variability of the errors, the rank order (i.e. relative performance) of the five methods stays the same as the testicular volume increases. The relevant *gllamm* command is given in Appendix 3(b).

The results are shown in Table 4.12(b). There appears to be a significant improvement in fit. If we now constrain the loadings to be equal we obtain a log-likelihood of −411.701. Twice the difference between this and that for the unconstrained model is 26.01 with 4 d.f. There still appears to be a statistically significant difference between the slopes (βs) for the five methods. We could refine our model further. We could allow for the slope of the effect of logged mean volume on the logarithm of the standard deviation of the measurement errors to vary with measurement method. (The lines might even cross – the performance of a method relative to another being dependent on the true volume.) We might search for a better sub-model to explain the relationship between the variability of the measurement errors and true volume. And we could improve our approximation of a subject's true volume by estimating the true scores from our first fitted model (using a function of the observed subject means), and letting variability of the measurement errors be dependent upon the same function of the estimated true values in a re-fitted model (with the possibility of

Table 4.12 Results of fitting measurement models to testicle volume data in *gllamm*.

(a) The basic model – homoscedastic errors within methods

Method	Intercept, α (s.e.)	Slope, β (s.e.)	log SD of errors (s.e.)
US	0*	1*	0.147 (0.139)
G	0.099 (0.567)	0.884 (0.051)	0.144 (0.131)
D	−0.012 (0.623)	0.950 (0.056)	0.260 (0.129)
P	0.039 (0.627)	1.143 (0.056)	0.065 (0.161)
R	1.535 (0.635)	1.083 (0.057)	0.174 (0.141)
γ^{**}	0		
μ_T	10.081 (0.749)		
σ_T^2	22.235 (5.142)		
$\log L$	−424.774		

* Constraint.
** The slope of the line regressing the logarithm of the standard deviation of the measurement errors on the centred logarithm of the subject's mean volume measurement – here constrained to be 0.

(b) The revised model – heteroscedastic errors within methods

Method	Intercept, α (s.e.)	Slope, β (s.e.)	log SD of errors (s.e.)
US	0*	1*	0.057 (0.140)
G	0.266 (0.407)	0.867 (0.048)	0.079 (0.130)
D	0.094 (0.411)	0.939 (0.049)	0.024 (0.142)
P	0.122 (0.471)	1.135 (0.056)	0.101 (0.151)
R	1.744 (0.468)	1.062 (0.056)	0.161 (0.138)
γ^{**}	0.617 (0.114)		
μ_T	10.051 (0.753)		
σ_T^2	22.511 (5.201)		
$\log L$	−411.701		

* Constraint.
** The slope of the line regressing the logarithm of the standard deviation of the measurement errors on the centred logarithm of the subject's mean volume measurement.

further iterations). We will not pursue this problem here, however. It was introduced as a simple heuristic device to illustrate how we might make our models more realistic. Readers wishing to develop this theme are referred to the papers by Meijer and Mooijaart and by Lewin-Koh and Amemiya, cited above. Other approaches to the joint modeling of means and error variances (with particular reference to the calibration of chemical assay data) are described in Rocke and Lorenzato (1995), Zorn *et al.* (1997), and in Gibbons and Bhaumik (2001).

4.9 Replication

Returning to the original identifiability problem for the comparison of two measurement instruments, we have seen that this can be solved by finding an appropriate instrumental variable which may or may not be a third measuring instrument. A further way of solving the problem is to repeat the measurements using one or both of the original two instruments (that is, to use a replicate measurement as an instrumental variable). In all cases we assume that the measurements are conditionally independent given the true value of the characteristic being measured. That is, we assume that there are no correlated measurement errors. Correlations between measurement errors are *always* a potential pitfall, but they are particularly likely when we repeat the measurement process using the same method. Any subject-specific bias in the use of a given method is going to be common to the replications of the measurements on that subject. Such correlated measurement errors need to be detected and allowed for in the statistical analysis. Otherwise we will obtain biased estimates of our model parameters and potentially come to the wrong conclusions. Ideally, we should have a design which will enable us to detect correlated measurement errors and to deal with them appropriately. But this will not always be possible. In this section we introduce models for data involving repeated measurements using the same instrument(s) in order to clarify some of these ideas.

This section deals with the more general case of introducing replicated measurements, whether or not they are actually needed for identifiability. When dealing with replication we first need to distinguish two types of data structure. In most of the examples discussed below we will be wishing to compare measurements resulting from the use of two or more methods (X, Y, Z, etc) in which we have, say, a pair of measurements taken at the first time point (X_1 and Y_1) followed at a later time by a second (X_2 and Y_2) and subsequent pairs of measurements. The characteristic being measured may or may not have changed in the interval between the successive measurement occasions. In this section we will assume that the characteristic being measured is stable (the situation in which it changes over time will be dealt with in Section 4.15). The important point to bear in mind, however, is that X_1 is always paired with Y_1, X_2 with Y_2, and so on. There is a particular structure determined by the design of the experiment and this structure should be accounted for in the appropriate measurement models and accompanying statistical inference. That is, the repeated measurements using a given instrument are not exchangeable and should not inadvertently be analyzed using a statistical model based on the assumption that they are. If we are analyzing data we have collected ourselves then the structure of these data should be clear. If, however, we are analyzing someone else's data (as for many of the examples in the present book) it is not always obvious how the data have been collected. The design leading to a given data set is frequently not described with sufficient precision, and we are usually making inferences concerning the design from the way the resulting data have been tabulated. Replicate measurements which are exchangeable arise when we take, for example, five measurements using method X and another five using method Y. Here there is no link between any particular X measurement with any one of the Ys, and there is no ordering of either the Xs or of the Ys. But if there is a distinct ordering and, for example, the sequence of Xs drift over time (irrespective of or in addition to what might be happening to the Ys), this should then be taken into account. Observers using subjective methods may be improving with practice (or getting worse through drifting away from specified operational criteria). In a laboratory there may be instrumental drift, or deterioration of assay materials, or changes in the skill of the person using the instrument or assay, or all of these. There may be serial correlations between the measurement errors, and so on. We will start with examples of models using measurements which are not exchangeable (or at least, assumed not to be) and will then introduce examples with exchangeable data at the end of the section. Before doing this, however, it might be useful to describe a few data sets to set the scene.

Oman *et al.* (1999) describe a study to compare two methods of measuring creatinine clearance (an index of kidney function) in cancer patients undergoing chemotherapy with a potentially nephrotoxic drug – the standard method, MCC, and a new and easier method,

ECC. Their description of the resulting data indicate that the MCC and ECC measurements are clearly paired:

> Our data comprise repeated observations during the course of treatment of 110 consecutive patients who underwent cisplatin chemotherapy in the Department of Oncology at Hadassah Hebrew University Hospital. MCC and ECC were routinely determined before the first course of cisplatin, and thereafter from 10 to 14 days following each cisplatin administration.

In this example, however, the characteristic being measured may be varying over time. We will return to this type of data in Section 4.15.

A particularly well-known data set is the one presented and analyzed in Bland and Altman (1986) and several subsequent publications by these authors. Here two measurements of lung function (Peak Expiratory Flow Rate – PEFR) were made on each patient using a Wright peak flow meter and two with a Mini Wright meter. We assume that the measurements for each patient within each of the two methods are exchangeable. Summary statistics for these PEFR data are given in Table 4.13. Bland and Altman (1999) describe a data set with a very similar structure to that of Oman *et al.* (1999), in which two methods of measuring cardiac output, impedence cardiography (IC) and radionuclide ventriculography (RV), were each used at regular intervals during surgery. They assume that there is no change in cardiac output (or instrumental drift) over time and analyze the data as if the replicates are exchangeable.

The same authors (Bland and Altman, 1999, Table 1) also present a large data set on systolic blood pressure measurements made simultaneously by two observers (*J* and *R*) and an automatic blood pressure measuring machine (*S*). Each of these three simultaneous sets of measurements were taken at three times in quick succession. Clearly this is a highly structured data set and the replicate measurements should not be considered as being exchangeable. These data could be analyzed using the methods for longitudinal data described in Section 4.15. Batista-Foguet *et al.* (2001) describe a survey of 5855 patients in which each patient had his or her blood pressure measured six times using a mercury sphygmomanometer. Both diastolic and systolic blood pressure was measured on the patient in a seated position immediately after arrival at the doctor's surgery. These two measurements were then both made a few minutes later. Finally, both measurements were taken on the patient in a standing position a few minutes after the second readings. Again this provides a data set with a relatively complex structure and it too will be discussed in Section 4.14. Finally, we mention a further set of blood pressure measurements in which the performance

Table 4.13 Summary statistics for the Bland and Altman lung capacity (PEFR) data ($N = 17$).

Variable	Mean	Standard deviation
Wright1	450.3529	116.3126
Wright2	445.4118	119.6129
Mini Wright1	452.4706	113.1151
Mini Wright 2	455.3529	111.3249

Correlations

	Wright1	Wright2	Mini1	Mini2
Wright1	1.0000			
Wright2	0.9838	1.0000		
Mini1	0.9433	0.9571	1.0000	
Mini2	0.9363	0.9525	0.9670	1.0000

of three methods of measurement (two automatic methods being compared to the standard mercury sphygmomanometer) in a three period crossover design (Bassein *et al.*, 1985).

In Chapter 3 we described the measurement of ventricle-brain ratios (VBRs) and presented data on a sample of 50 schizophrenic patients (Table 3.11, taken from Dunn, 1989). In this table there are repeated measurements of the logarithm of the VBR using each of two methods of measurement (a hand held planimeter and a computerized pixel count). It is not clear whether the within-method replicates (PIX1 and PIX3, or PLAN1 and PLAN3) are exchangeable but it seems to be a reasonable assumption that this is so. These measurements, however, were taken on the largest 'slice' of the head from a three-dimensional CAT scan (Turner *et al.*, 1986). Other measurements (PIX2 and PLAN2) were also taken on the second largest 'slice'. Clearly neither PIX1 and PIX2 nor PLAN1 and PLAN2 are exchangeable. They are also measuring a characteristic (the logarithm of the VBR indicated by the 2nd slice) which may be slightly different from, but presumably very highly correlated with, the logarithm of the VBR as indicated by measurements from the first slice. Strike's gentamicin assay data (also discussed in Chapter 3) is a similar example.

> Serum specimens from fifty-six patients receiving gentamicin have been assayed twice by each assay method in separate assay batches. (Strike, 1995, page 151).

Consider the data given in Table 4.14. These measurements of wood veneer quality are taken from an experiment reported in Kauman *et al.* (1956). These authors provide a description of their experimental procedure as follows:

> Twenty sheets of 1/16 in. thick veneer (green size 38 × 38 in.) from five trees of alpine ash (*Eucalyptus gigantea* Hook f.) were selected from material included in a study of the mechanical drying of 'ash' type eucalypt veneer. The sheets had been selected to include, as far as possible, all types and severities of degrade and quality ratings.
>
> Three observers (*A*, *B* and *C*) each carried out two quality evaluations on these twenty veneer sheets, using first the subjective and then the numerical scheme. Several days were allowed to pass between the first and the second evaluation by each observer, and the order of the sheets was changed between tests. No observer was permitted to see the others' results, or to be present during the others' tests, before he had completed all his own tests.

Here there may be consistent differences between the Observers' ratings at the first and second occasions. They may not be exchangeable, although there is no reason to believe that the first rating by Observer *A*, for example, should be linked to the first ratings of either *B* or *C*. We could start by using a model which does not assume that the 1st and 2nd measurement occasions are equivalent, and use the data to test whether the equivalence (exchangeability) assumption might be justified. These data have also been analysed in some detail in the book by Mandel (1991) who takes a very different (but complementary) approach to the one described here. We briefly referred to the analysis of these data in Chapter 3 (see Section 3.8).

Mandel (1991) also analyses data from an inter-laboratory precision study for the measurement of pentose sugars (pentosans) in wood pulp. In this case the seven different analytical laboratories are analogous to seven different measuring instruments. The data are shown in Table 4.15. The within-laboratory replicates are clearly exchangeable. Permutt *et al.* (1991) describe the results of an inter-laboratory precision study from the National Stream Survey. In this study seven different audit samples (of water) were sent to each of two laboratories. Each of the seven samples was assayed for a large range of chemical components. There were replicate measurements taken by each laboratory on each audit sample, but, unlike the pentose sugar data, the number of replicates varied between samples within each

Table 4.14 Measurements of Eucalypti veneer quality.

Sheet	Observer A		Observer B		Observer C	
	Test 1	Test 2	Test 1	Test 2	Test 1	Test 2
1	22	15	21	24	19	11
2	31	23	24	21	29	33
3	34	37	32	35	40	37
4	44	44	41	44	42	40
5	10	6	13	19	8	7
6	31	32	34	39	34	25
7	42	39	39	41	25	30
8	28	19	23	27	13	24
9	29	29	27	27	28	25
10	39	37	29	35	34	36
11	21	20	17	25	16	23
12	42	44	44	40	41	38
13	40	39	34	37	34	31
14	11	9	9	12	7	8
15	38	37	30	36	44	40
16	9	13	14	25	11	12
17	20	16	19	19	6	15
18	28	31	30	31	24	37
19	28	24	25	30	25	22
20	15	8	11	19	9	10

Source: Kauman *et al.* (1956).

Table 4.15 Pentose sugar concentrations in pulp.

Laboratory	Material								
	A	B	C	D	E	F	G	H	I
1	0.44	0.96	1.23	1.25	1.98	4.12	5.94	10.70	17.13
	0.49	0.92	1.88	1.25	1.92	4.16	5.37	10.74	16.56
	0.44	0.82	1.24	1.42	1.80	4.16	5.37	10.83	16.56
2	0.41	0.83	1.12	1.25	1.99	4.10	5.26	10.07	16.08
	0.41	0.83	1.12	1.25	1.94	4.11	5.26	10.05	16.04
	0.41	0.84	1.12	1.26	1.95	4.10	5.26	9.82	16.13
3	0.51	0.92	1.11	1.35	2.05	4.11	5.16	10.01	16.01
	0.51	0.93	1.13	1.35	2.08	4.16	5.16	10.17	15.96
	0.51	0.92	1.11	1.35	2.03	4.16	5.21	10.17	16.06
4	0.40	0.96	1.15	1.29	2.05	4.20	5.20	10.98	16.65
	0.38	0.94	1.13	1.29	2.04	4.20	5.20	10.67	16.91
	0.37	0.94	1.13	1.29	2.04	4.22	5.20	10.52	16.75
5	0.49	0.82	0.98	1.23	1.94	4.61	5.00	10.48	15.71
	0.49	0.82	0.98	1.23	1.96	4.63	5.00	10.27	15.45
	0.49	0.84	0.98	1.23	1.96	4.53	4.96	10.38	15.66
6	0.43	0.88	1.11	1.31	2.01	3.93	4.85	9.57	15.05
	0.41	0.92	1.12	1.30	1.99	3.92	4.87	9.57	14.73
	0.40	0.88	1.11	1.31	1.98	3.84	4.91	9.62	15.04
7	0.186	0.866	1.05	1.13	1.98	4.21	5.27	11.5	18.8
	0.171	0.900	0.962	1.15	1.93	4.18	5.32	10.8	18.2
	0.153	0.831	0.927	1.15	1.98	4.16	5.10	11.5	18.1

Source: Mandel (1991).

Table 4.16 Summary statistics for Kauman's veneer quality data ($N = 20$).

Variable	Mean	Standard deviation
A1	28.1	11.145
A2	26.1	12.397
B1	25.8	10.050
B2	29.3	8.874
C1	24.45	12.734
C2	25.2	11.400

Correlations

	A1	A2	B1	B2	C1	C2
A1	1.000					
A2	0.954	1.000				
B1	0.937	0.957	1.000			
B2	0.876	0.927	0.933	1.000		
C1	0.876	0.914	0.857	0.841	1.000	
C2	0.896	0.919	0.844	0.794	0.885	1.000

of the two laboratories. Again, the within-sample replicates within each of the two laboratories are exchangeable. Inter-laboratory precision studies are the subject of Section 4.11.

We start our analyses by fitting a straightforward measurement model to the PEFR data. In their discussions of the analyses of these data, Bland and Altman implicitly assume that the two Wright meter PEFR measurements and the two provided by the Mini Wright meter have no fixed or relative biases with respect to one another (i.e. they are tau-equivalent). Their estimates of the repeatability of the two methods are also dependent on the assumption of the conditional independence of the measurement errors. These assumptions lead to the following model:

$$WRIGHT1_i = \tau_i + \omega_i$$
$$WRIGHT2_i = \tau_i + \delta_i$$
$$MINI1_i = \tau_i + \varepsilon_i \tag{4.18}$$
$$MINI2_i = \tau_i + \upsilon_i$$

with both $Var(\omega_i) = Var(\delta_i)$ and $Var(\varepsilon_i) = Var(\upsilon_i)$. Fitting this model produces a goodness-of-fit chi-square statistic of 12.611 with 10 d.f. (P = 0.246). The model fits very well (but bear in mind the small sample size). The estimate of the variance of the Wright meter measurement errors is 310.592 (s.e. 107.601). The corresponding estimate for the Mini Wright meter is 969.137 (s.e. 275.565). The variance of τ is estimated to be 13138.032 (s.e. 4686.677) so that the reliabilities of the Wright meter and the Mini Wright meter are 0.977 and 0.931, respectively. The estimate of the variance of the difference between single Wright and Mini Wright readings on individual subjects is 1279.729 (969.137 + 310.592). The corresponding standard deviation is therefore 35.773. Using the first measurements only for the Wright and Mini Wright meters, Bland and Altman produced an estimate of 38.8. Either of these can be used to calculated limits of agreement as described earlier.

Now let's have a look at the veneer data of Kauman *et al.* (summary statistics are given in Table 4.16). If we fit the equivalent model to (4.18), but modifying it to take account of the fact that we have three measurements, we get a chi-square of 57.606 with 22 d.f.

($P < 0.001$). This is a particularly badly fitting model. We now relax a few of the constraints and try fitting a model in which Observers B and C have non-zero intercepts and non-unit slopes (allowing these to differ between Observers but constraining them to be equal for the replicates within Observers). We still constrain the within-Observer estimates of measurement error variance to be the same for the replicates. The model still does not fit! Chi-square equals 39.199 with 18 d.f. ($P < 0.003$). The results of fitting a model with no constraints is shown in Table 4.17(a). Chi-square for this model is 12.407 with 9 d.f. This fits and also gives us a check that there are no significant departures from the assumption of conditional independence of the measurement errors. Exploring all of the obvious options between fitting the unconstrained model and the second one above leads us to conclude that the model in Table 4.17(b) provides a reasonable (but not entirely satisfactory) description of the data (chi-square = 23.643 with 15 d.f.; $P = 0.071$). It appears that there are within-Observer shifts in the means (i.e. the intercept terms) from the first replicate to the second. We have no explanation for this effect. Note that the measurement error variances increase as we

Table 4.17 Results of fitting an unconstrained measurement model (with conditional independence) to the veneer data.

(a) No constraints

Parameter	Estimate (s.e.)	Parameter	Estimate (s.e.)
α_{A1}	0*	β_{A1}	1*
α_{A2}	−6.420 (2.485)	β_{A2}	1.157 (0.083)
α_{B1}	0.409 (2.603)	β_{B1}	0.904 (0.087)
α_{B2}	7.612 (2.657)	β_{B2}	0.772 (0.088)
α_{C1}	−6.241 (4.001)	β_{C1}	1.092 (0.133)
α_{C2}	−2.365 (3.542)	β_{C2}	0.981 (0.118)
σ^2_{A1}	10.153 (3.600)		
σ^2_{A2}	0.925 (1.681)		
σ^2_{B1}	7.879 (2.813)		
σ^2_{B2}	10.802 (3.651)		
σ^2_{C1}	26.094 (8.746)		
σ^2_{C2}	20.181 (6.776)		
μ_T	28.100 (2.557)	σ^2_T	114.058 (40.191)

(b) Slopes and error variances equal within Observers

Parameter	Estimate (s.e.)	Parameter	Estimate (s.e.)
α_{A1}	0*	β_{A1}	1*
α_{A2}	−2.000 (0.827)	β_{A2}	1*
α_{B1}	3.937 (1.704)	β_{B1}	0.778 (0.053)
α_{B2}	7.437 (1.704)	β_{B2}	0.778 (0.053)
α_{C1}	−2.517 (2.527)	β_{C1}	0.960 (0.078)
α_{C2}	−1.767 (2.527)	β_{C2}	0.960 (0.078)
σ^2_{A1}	6.849 (1.974)		
σ^2_{A2}	6.849 (1.974)		
σ^2_{B1}	9.691 (2.525)		
σ^2_{B2}	9.691 (2.525)		
σ^2_{C1}	24.050 (5.959)		
σ^2_{C2}	24.050 (5.959)		
μ_T	28.100 (2.704)	σ^2_T	132.459 (44.035)

pass from Observer A, to Observer B to Observer C. This can be explained by the following quotation from Kauman *et al.*:

> Observer A was most experienced, having completed a quality evaluation of some 300 sheets of veneer by both schemes immediately prior to the present experiment; B had acted as a recorder to A during this evaluation, but had not himself carried out any of the evaluations, whereas C had no previous knowledge of the schemes and carried out his evaluation after a brief training period.

Moving on to the CAT scan measurements, we start with the following model:

$$
\begin{aligned}
PLAN1_i &= \tau_i + \omega_i \\
PLAN3_i &= \alpha_{PLAN3} + \beta_{PLAN3}\tau_i + \delta_i \\
PIX1_i &= \alpha_{PIX1} + \beta_{PIX1}\tau_i + \varepsilon_i \\
PIX3_i &= \alpha_{PIX3} + \beta_{PIX3}\tau_I + \upsilon_i
\end{aligned}
\tag{4.19}
$$

where the notation should be clear from the context. Despite the lack of constraints, the model does not fit (chi-square = 28.743 with 2 d.f.; P < 0.001). If we introduce the obvious 6 constraints $\alpha_{PLAN3} = 0$, $\beta_{PLAN3} = 1$, $\alpha_{PIX1} = \alpha_{PIX3}$, $\beta_{PIX1} = \beta_{PIX3}$, $Var(\omega_i) = Var(\delta_i)$ and $Var(\varepsilon_i) = Var(\upsilon_i)$ we get a chi-square of 38.327 with 8 d.f., which is not significantly worse than before (chi-square for the difference is 9.584 with 6 d.f.). Looking at the raw data indicates there are several pairs of Pixel measurements which are both considerably less than the corresponding Planimetry measures (one obvious example is where both Pixel measurements are 0). It appears that although the Pixel method is very consistent it might occasionally lead to gross error, and this error is common to both replicates. On the other hand, although they are more erratic, the Planimeter-based measurements appear to be much less prone to these gross errors. We allow for this by letting the two Pixel measurement errors (that is, ω_i and δ_i) be correlated. The change in fit is striking (chi-square = 2.502 with 1 d.f.). Table 4.18(a) provides parameter estimates for the first model; Table 4.18(b) gives those for the same model but with the added correlation between the two measurement errors. It is hard to believe that these two sets of estimates come from the same data! In particular, note the changes in the slopes for the Pixel counts and the measurement error variances for all four methods.

Starting with the unconstrained model with correlated measurement errors for the Pixel counts, we first constrain the variances of the measurement errors to be the same within the two methods (that is, $Var(\omega_i) = Var(\delta_i)$ and $Var(\varepsilon_i) = Var(\upsilon_i)$) to obtain a chi-square of 6.724 with 3 d.f. The parameter estimates for this model are given in Table 4.18(c). In addition, we could introduce constraints on the intercepts ($\alpha_{PLAN3} = 0$, $\alpha_{PIX1} = \alpha_{PIX3}$), leading to a chi-square of 10.306 with 5 d.f., or on the slopes ($\beta_{PLAN3} = 1$, $\beta_{PIX1} = \beta_{PIX3}$), leading to a chi-square of 8.271, again with 5 d.f. If, however, we try to constrain both intercepts and slopes, we get a model that does not fit (chi-square = 22.738 with 7 d.f.; P < 0.002). We finally fit a model in which both intercepts and slopes are constrained, but for the Pixel measures only (chi-square = 8.678 with 5 d.f.). The parameter estimates for this model are given in Table 4.18(d).

Before leaving the CAT scan data, let's assume that we have not repeated the Pixel measure. So, our data now comprises measurements on PLAN1, PLAN3 and PIX1. We wish to fit a measurement model for PLAN1 and PIX1 using PLAN3 as an instrumental variable. The results of this exercise are presented in Table 4.19(a). As might be expected from the

Table 4.18 Results of fitting an unconstrained measurement model to the CAT scan data.

(a) No constraints, but errors for the pixel counts assumed to be uncorrelated

Parameter	Estimate (s.e.)	Parameter	Estimate (s.e.)
α_{PLAN1}	0*	β_{PLAN1}	1*
α_{PLAN3}	−0.800 (0.596)	β_{PLAN3}	1.349 (0.318)
α_{PIX1}	−2.667 (0.801)	β_{PIX1}	2.186 (0.427)
α_{PIX3}	−2.676 (0.805)	β_{PIX3}	2.197 (0.429)
σ^2_{PLAN1}	0.105 (0.021)		
σ^2_{PLAN3}	0.088 (0.018)		
σ^2_{PIX1}	0.001 (0.003)		
σ^2_{PIX3}	0.002 (0.003)		
μ_T	1.862 (0.058)	σ^2_T	0.057 (0.025)

(b) No constraints, but errors for the pixel counts assumed to be correlated

Parameter	Estimate (s.e.)	Parameter	Estimate (s.e.)
α_{PLAN1}	0*	β_{PLAN1}	1*
α_{PLAN3}	−0.854 (0.395)	β_{PLAN3}	1.378 (0.210)
α_{PIX1}	−0.856 (0.396)	β_{PIX1}	1.213 (0.210)
α_{PIX3}	−0.837 (0.400)	β_{PIX3}	1.209 (0.212)
σ^2_{PLAN1}	0.062 (0.016)		
σ^2_{PLAN3}	0.001 (0.020)		
σ^2_{PIX1}	0.125 (0.029)		
σ^2_{PIX3}	0.129 (0.030)		
μ_T	1.862 (0.058)	σ^2_T	0.100 (0.032)
		$\sigma_{\delta\omega}$	0.125* (0.030)

* Equivalent to an estimated correlation of 0.988.

(c) Constraints on error variances, and errors for the pixel counts assumed to be correlated

Parameter	Estimate (s.e.)	Parameter	Estimate (s.e.)
α_{PLAN1}	0*	β_{PLAN1}	1*
α_{PLAN3}	−0.418 (0.243)	β_{PLAN3}	1.378 (0.210)
α_{PIX1}	−0.705 (0.346)	β_{PIX1}	1.132 (0.183)
α_{PIX3}	−0.709 (0.347)	β_{PIX3}	1.140 (0.183)
σ^2_{PLAN1}	0.038 (0.008)		
σ^2_{PLAN3}	0.038 (0.008)		
σ^2_{PIX1}	0.119 (0.028)		
σ^2_{PIX3}	0.119 (0.028)		
μ_T	1.862 (0.057)	σ^2_T	0.120 (0.032)
		$\sigma_{\delta\omega}$	0.117* (0.030)

* Equivalent to an estimated correlation of 0.987.

(d) Constraints on error variances, and errors for the pixel counts assumed to be correlated. Intercepts and slopes for pixel counts also constrained

Parameter	Estimate (s.e.)	Parameter	Estimate (s.e.)
α_{PLAN1}	0*	β_{PLAN1}	1*
α_{PLAN3}	−0.419 (0.243)	β_{PLAN3}	1.144 (0.129)
α_{PIX1}	−0.708 (0.346)	β_{PIX1}	1.136 (0.183)
α_{PIX3}	−0.708 (0.346)	β_{PIX3}	1.136 (0.183)
σ^2_{PLAN1}	0.038 (0.008)		
σ^2_{PLAN3}	0.038 (0.008)		
σ^2_{PIX1}	0.119 (0.028)		
σ^2_{PIX3}	0.119 (0.028)		
μ_T	1.862 (0.057)	σ^2_T	0.120 (0.032)
		$\sigma_{\delta\omega}$	0.117* (0.028)

* Equivalent to an estimated correlation of 0.986.

Table 4.19 Results of fitting an unconstrained measurement model to the CAT scan data: using the repeat measurement from only one of the two methods.

(a) PLAN1 and PIX1, with PLAN3 as the instrumental variable

Parameter	Estimate (s.e.)	Parameter	Estimate (s.e.)
α_{PLAN1}	0*	β_{PLAN1}	1*
α_{PIX1}	−0.856 (0.394)	β_{PIX1}	1.213 (0.208)
α_{PLAN3}	−0.827 (0.387)	β_{PLAN3}	1.364 (0.206)
σ^2_{PLAN1}	0.061 (0.016)		
σ^2_{PIX1}	0.123 (0.029)		
σ^2_{PLAN3}	0.003 (0.019)		
μ_T	1.862 (0.058)	σ^2_T	0.101 (0.032)

(b) PLAN1 and PIX1, with PIX3 as the instrumental variable

Parameter	Estimate (s.e.)	Parameter	Estimate (s.e.)
α_{PLAN1}	0*	β_{PLAN1}	1*
α_{PIX1}	−2.636 (0.790)	β_{PIX1}	2.171 (0.421)
α_{PIX3}	−2.675 (0.797)	β_{PIX3}	2.196 (0.425)
σ^2_{PLAN1}	0.105 (0.021)		
σ^2_{PIX1}	0.003 (0.001)		
σ^2_{PIX3}	0.000 (3.567)		
μ_T	1.862 (0.058)	σ^2_T	0.057 (0.025)

very high correlations between the measurement errors for PIX1 and PIX3 found earlier, the results are essentially identical to those in Table 4.18(d). Now let's assume that the investigator took the easiest route to get an instrumental variable and repeated the Pixel count instead of the laborious Planimetry measure. We now have a data set comprising measurements on PLAN1, PIX1 and PIX3. Fitting our model to these three measurements produces the results in Table 4.19(b). Apart from the technical problem of an estimated error variance of zero, you will see that the parameter estimates are back to those we started with. They are invalid. However, if these measures were the only ones we have access to, then we would never know.

4.10 Equation Errors (subject-specific biases or random matrix effects)

Staying with the CAT scan measurements, we start with the following model:

$$
\begin{aligned}
PLAN1_i &= \tau_i + \omega_i \\
PLAN3_i &= \alpha_{PLAN3} + \beta_{PLAN3}\tau_i + \delta_i \\
PIX1_i &= \alpha_{PIX1} + \beta_{PIX1}\tau_i + \varepsilon_i \\
PIX3_i &= \alpha_{PIX3} + \beta_{PIX3}\tau_I + \upsilon_i
\end{aligned}
\tag{4.20}
$$

in which ω_i and δ_I are correlated. An entirely equivalent model (giving exactly the same chi-square and d.f.) is the following:

$$
\begin{aligned}
PLAN1_i &= \tau_i + \omega_i \\
PLAN3_i &= \alpha_{PLAN3} + \beta_{PLAN3}\tau_i + \delta_i \\
PIX1_i &= \alpha_{PIX1} + \beta_{PIX1}\tau_i + \gamma_i + \varepsilon_i \\
PIX3_i &= \alpha_{PIX3} + \beta_{PIX3}\tau_I + \gamma_i + \upsilon_i
\end{aligned}
\tag{4.21}
$$

where, in this case, we assume all errors are independent. The γs in model (4.21) are values of a latent variable being tapped by both the Pixel counts but not the Planimeter measures. They are uncorrelated with the τs. Instead of trying to interpret what the γs might signify, however, we will move on to analytical chemistry where they can frequently be interpreted as random matrix effects.

Consider, for example, a situation where, for the ith subject or specimen, we have two measures using the standard assay (X_{1i} and X_{2i}) together with two measures using the new technique (Y_{1i} and Y_{2i}). This is the situation for Strike's gentamicin data considered in Chapter 3. We assume that the replicates within each method are exchangeable. A possible model for these data is now given by

$$
\begin{aligned}
X_{1i} &= \tau_i + \delta_{1i} \\
X_{2i} &= \tau_i + \delta_{2i} \\
Y_{1i} &= \alpha + \beta\tau_i + \varepsilon_{1i} \\
Y_{2i} &= \alpha + \beta\tau_i + \varepsilon_{2i}
\end{aligned}
\tag{4.22}
$$

where $\mathrm{Var}(\delta_{1i}) = \mathrm{Var}(\delta_{2i}) = \sigma_\delta^2$ and $\mathrm{Var}(\varepsilon_{1i}) = \mathrm{Var}(\varepsilon_{2i}) = \sigma_\varepsilon^2$. Fitting this model provides the following estimates: $\beta = 0.956$ (s.e. 0.028) and $\alpha = 0.338$ (s.e. 0.204). σ_δ^2 and σ_ε^2 are estimated to be 1.587 (s.e. 0.229) and 0.213 (s.e. 0.040), respectively. Unfortunately the goodness of fit statistic (a chi-square of 38.02 with 8 degrees of freedom; $P < 0.001$) implies that there are problems with this model. There may be random matrix effects. A matrix of scatter plots (Figure 4.5) shows that there is at least one outlier, and also appears to indicate that there is a floor effect for the EMIT measurements – see bottom left corner of the EMIT vs FIA plots. It is also clear that the measurements are all positively skewed. We will ignore all of this, however.

First, let us simplify our notation in order to make our manipulations of variances and covariances less cumbersome. We will drop the subscript i referring to the observations on the ith subject. We will assume that we have four measurements on each of n subjects and label them as A, B, C and D. A and B are replicates using instrument or method X. C and D are replicates using Y. A and B have a random component in common which is not found

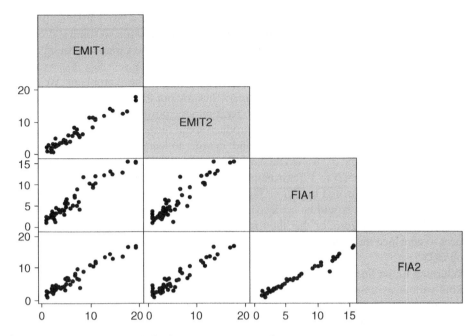

Figure 4.5 Scatterplot matrix for four measurements of gentamicin.

in C and D, and *vice versa*. These two random components (δ_x and δ_y, with variances $\sigma_{\delta x}^2$ and $\sigma_{\delta y}^2$, respectively) represent the postulated random matrix effects. The random errors that are unique to A, B, C and D are all mutually independent and labelled as ε_A, ε_B, ε_C and ε_D, respectively, with $\mathrm{Var}(\varepsilon_A) = \mathrm{Var}(\varepsilon_B) = \sigma_{\varepsilon x}^2$, and $\mathrm{Var}(\varepsilon_C) = \mathrm{Var}(\varepsilon_D) = \sigma_{\varepsilon y}^2$. In order to simplify matters even further, we will also forget about estimation of, or imposition of constraints on, the intercept term, α. Our model is the following:

$$
\begin{aligned}
A &= \tau + \delta_x + \varepsilon_A \\
B &= \tau + \delta_x + \varepsilon_B \\
C &= \alpha + \beta\tau + \delta_y + \varepsilon_C \\
D &= \alpha + \beta\tau + \delta_y + \varepsilon_D
\end{aligned}
\tag{4.23}
$$

If $\mathrm{Var}(\tau) = \sigma_\tau^2$ then the expected covariance matrix has the following structure:

$$
\begin{bmatrix}
\sigma_\tau^2 + \sigma_{\delta x}^2 + \sigma_{\varepsilon x}^2 & & & \\
\sigma_\tau^2 + \sigma_{\delta x}^2 & \sigma_\tau^2 + \sigma_{\delta x}^2 + \sigma_{\varepsilon x}^2 & & \\
\beta^2\sigma_\tau^2 & \beta^2\sigma_\tau^2 & \beta^2\sigma_\tau^2 + \sigma_{\delta y}^2 + \sigma_{\varepsilon y}^2 & \\
\beta^2\sigma_\tau^2 & \beta^2\sigma_\tau^2 & \beta^2\sigma_\tau^2 + \sigma_{\delta y}^2 & \beta^2\sigma_\tau^2 + \sigma_{\delta y}^2 + \sigma_{\varepsilon y}^2
\end{bmatrix}
\tag{4.24}
$$

If we now represent the observed (estimated) variances by S_{AA}, S_{BB}, and so on, and the corresponding covariances by S_{AB}, etc., then we can equate these values with their expectations and attempt to solve the resulting simultaneous equations. This we will now do in detail to illustrate the problems of identifiability. There are two possible estimates for $\sigma_{\varepsilon x}^2$. These are $S_{AA} - S_{AB}$ and $S_{BB} - S_{AB}$. Similarly, $\sigma_{\varepsilon y}^2$ is estimated by $S_{CC} - S_{CD}$ and $S_{DD} - S_{CD}$.

This is what we mean by over-identification in the context of these two particular parameters. There is more than one possible estimate and we therefore resort to some sort of optimization routine to find the 'best' one for the data as a whole. The variance term $\sigma_{\delta y}^2$ is also over-identified. There are four possible estimates: $S_{AC} - S_{CD}, S_{AD} - S_{CD}, S_{BC} - S_{CD}$ or $S_{BD} - S_{CD}$. But, for the remaining parameters (β, σ_τ^2 and $\sigma_{\delta x}^2$) we have problems. We cannot find any unique estimate for any of these three – they are not identified. Consequently the model as a whole is under-identified. We are back into the position we were in Chapter 3! In order to estimate β and σ_τ^2 we need to introduce some further assumptions concerning either or both of $\sigma_{\delta x}^2$ and $\sigma_{\delta y}^2$. On the other hand, in order to learn something about the relative magnitudes of $\sigma_{\delta x}^2$ and $\sigma_{\delta y}^2$ we would have to introduce an assumption concerning either β or σ_τ^2. If we assume that $\beta = 1$, for example, we can immediately estimate σ_τ^2 by S_{AC}, S_{AD}, S_{BC} and S_{BD}. $\sigma_{\delta x}^2$ is estimated by $S_{AA} - S_{AB}$ and $S_{BB} - S_{AB}$. The model is now identified. Similarly, identification would be achieved by assuming that either of the random matrix effects variances were zero, or that their ratio was fixed at a known value. We leave the readers to convince themselves that this is true. We will illustrate the effects of doing this using *EQS*.

Returning to the Strike data on gentamicin determination, let us try to estimate β after making various assumptions concerning either $\sigma_{\delta x}^2$ or $\sigma_{\delta y}^2$, or both. We will fit the various models using maximum likelihood in *EQS*. First, if we assume that $\sigma_{\delta y}^2$ is zero we obtain a chi-square goodness-of-fit statistic of 18.985 with 5 degrees of freedom (P = 0.002). The model does not fit very well, but we will ignore this for our present purposes. β is estimated to be 0.959 with a standard error of 0.035. If we now assume that $\sigma_{\delta x}^2$ is zero (rather than $\sigma_{\delta y}^2$) we obtain an identical goodness-of-fit and our estimate of β is now 0.917 with a standard error of 0.033. Now, if we assume that we know the relative magnitudes of $\sigma_{\delta x}^2$ and $\sigma_{\delta y}^2$ (assuming, for example, that they are equal) then, again, the chi-square is 18.985 with 5 d.f. The estimate of β is 0.936 (s.e. 0.034). Note that the estimate of σ_τ^2 is also fluctuating with the changes in β. If we constrain both random matrix variances to be zero then we get a chi-square of 35.613 with 6 d.f. It is clear that there are significant random matrix effects, but we have insufficient data (i.e. the wrong design) to estimate them effectively. Finally, if we constrain $\beta = 1$ and attempt to estimate both $\sigma_{\delta x}^2$ or $\sigma_{\delta y}^2$ then we obtain an improper estimate (0) for $\sigma_{\delta y}^2$! The results of fitting the various models with random matrix effects are summarized in full in Table 4.20.

If we have random matrix effects then we are faced, yet again, with an identifiability problem. If we only have access to two methods of measurement (irrespective of how many

Table 4.20 Results of fitting models containing random matrix effects to Strike's gentamicin measurements.

Constraints	Estimate (s.e.)			
Parameter	$\sigma_{\delta y}^2 = 0$	$\sigma_{\delta x}^2 = 0$	$\sigma_{\delta y}^2 = \sigma_{\delta x}^2$	$\beta = 1$
$\sigma_{\delta y}^2$	0*	0.77 (0.24)	0.41 (0.13)	0.00 (0.03)
$\sigma_{\delta x}^2$	0.87 (0.27)	0*	0.41 (0.13)	0.91 (0.30)
$\sigma_{\varepsilon y}^2$	0.20 (0.04)	0.20 (0.04)	0.20 (0.04)	0.20 (0.04)
$\sigma_{\varepsilon x}^2$	0.77 (0.15)	0.77 (0.15)	0.77 (0.15)	0.77 (0.15)
σ_τ^2	19.11 (3.88)	19.98 (3.89)	19.57 (3.89)	17.66 (3.39)
β	0.96 (0.04)	0.92 (0.03)	0.94 (0.03)	1*

*Constraint.

replicates we can obtain using each of them) we cannot simultaneously estimate all of the required variance terms together with the regression coefficient, β. If this is our aim, then we need a third method of measurement, with replication of all three.

One final point: if we have evidence of random matrix effects then we need to be careful how we might define precision and/or reliability. Considering method X, for example, is $\sigma_{\varepsilon x}^2$ the variance of the 'errors' or is it $\sigma_{\delta x}^2 + \sigma_{\varepsilon x}^2$? Is the reliability given by $(\sigma_\tau^2/(\sigma_\tau^2 + \sigma_{\delta x}^2 + \sigma_{\varepsilon x}^2)$ or by $(\sigma_\tau^2 + \sigma_{\delta x}^2)/(\sigma_\tau^2 + \sigma_{\delta x}^2 + \sigma_{\varepsilon x}^2)$? Of course, it depends on the context.

Before leaving the gentamicin data it would be useful if we were able to explain the lack of fit of our models. We have not made any constraints on the intercepts (in effect fitting our model to an observed covariance matrix) but have constrained the other parameters to be equal across replications within methods. Here let's assume that we have set the variance of the random matrix effects for the EMIT measurements to be zero. If we constrain the two intercept terms for the FIA measures to be the same, but release the constraints on both the intercepts and slopes for EMIT1 and EMIT2 we obtain a chi-square of 7.255 with 5 d.f. EMIT1 is the standard (so that $\alpha = 0$ and $\beta = 1$) and the corresponding estimates of α and β for EMIT2 are 0.310 (s.e. 0.231) and 0.888 (s.e. 0.030), respectively. The intercept term for EMIT2 is not significantly different from 0, but the slope is clearly lower than 1. I have no explanation for this unexpected behaviour of the second EMIT measurement. With hindsight it should have been obvious. A simple Deming regression of EMIT2 on EMIT1, assuming equal measurement error variances (i.e. $\lambda = 1$), yields a slope of 0.890 (s.e. 0.031).

4.11 Inter-laboratory precision studies

In general, p laboratories will test q materials, and perform n replicate tests on each material, The materials should be chosen so as to cover the entire range of interest, as shown by the example. Each material should be made as homogeneous as possible prior to distribution to the laboratories. A replicate (test result) consists in carrying out *all* the steps of the measuring process (not just part of it). Thus if the test is a chemical analysis consisting of a *wet chemistry* part, followed by an instrumental, physical measurement such as reading of an absorbance at a specified wavelength of the spectrum, each replicate should include both of these phases, not just the reading of absorbance.   (Mandel, 1991, page 170)

The example referred to in this paragraph is the study of pentose sugar measurements, the results of which are presented in Table 4.15. Here we have seven laboratories making triplicate measurements on each of nine samples of pulp. Note the spacing and range of the pentose measurements across these samples – it would probably not be safe to assume that the concentration of pentose sugars was normally distributed across these samples. Often the actual amount or concentration of the given substance is known. Gibbons and Bhaumik (2001), for example, describe a blind inter-laboratory study in which seven laboratories provide five replicate measurements on samples of five known concentrations of copper (0, 2, 10, 50 and 200 μg/L.). 'Blind' here means that the laboratories did not know the true concentrations of copper in these samples.

If two independent laboratories were each to measure a particular chemical component in samples from a variety of freshwater lakes, for example, we could then analyse the resulting data using methods similar to those described earlier in this chapter as they are entirely analogous to data obtained from a method comparison study. Unlike the majority of method comparison studies, however, there are typically replicate measurements taken on each sample by each of the laboratories. And although we have not covered this possibility for method comparison studies so far, it is also likely that the samples to be measured are not randomly selected from a population of interest but are specifically chosen such that the concentration of the chemical component under study spans the entire range of interest (as in the measurement of pentose sugars or copper in the two examples above). In this situation we might be interested in fitting a set of simultaneous linear functional relationships (see Section 3.4) rather than the more usual structural models described so far. It is also likely that in an inter-laboratory precision study, unlike method comparison studies, we will have a sample of the population of eligible laboratories and will be interested in evaluating the variability of the components of bias (i.e. the αs and βs) instead of estimating them for each laboratory individually. A final complication is that the variance of the measurement errors will be a function of the amount of material being measured – heteroscedastic errors – see Section 4.8).

Let's start with the pentose sugar measurements in Table 4.21. Looking at these data we can see that the within-laboratory variability increases with the amount of sugar being measured. We will fit a model allowing for this effect below, but first we take a logarithmic transformation of the data to stabilize this variance and then fit a simple measurement model to these transformed measurements (using, for this example, the *gllamm* command in *Stata*). Here we are analysing the data as if it were from a method comparison study, treating each laboratory as if it were a separate method of measurement. Ignoring the above warning about the concentrations not being normally distributed, we fit the standard structural measurement model assuming multivariate normality. The results are given in Table 4.21(a). The slopes (βs) are all fairly close to 1 but there is variability between laboratories. Note that Laboratory 7 seems to be a bit of an outlier: its intercept (α) and slope (β) are quite different to those of the other six laboratories, and the variability of its measurement errors (log SD of errors) also seems to be higher.

What if we wish to fit the corresponding functional model? Here, in addition to estimating the relevant 'measurement' parameters for each of the laboratories, we are interested in estimating the individual true means for each of the nine pulp samples, not their mean and variance as in the structural model. Well, a simple starting point might be to estimate the true concentration for each of the nine samples from the mean of the 21 (i.e. 7×3) measurements from that sample and then regress the measurements from each of the seven laboratories separately to get estimates of α, β and log SD of the errors. The results of using *gllamm* to carry out these estimations is given in Table 4.21(b). This is an approach which is quite often taken in the analysis of inter-laboratory precision studies (see Mandel, 1964 and 1991, for example). Qualitatively, the results are very similar to those already obtained. The main difference is that the standard errors of the parameter estimates are noticeably smaller, which is presumably a reflection of the fact that the nine sample means are not being acknowledged as fallible estimates but are being treated, incorrectly, as if they were known true values.

Now let's sketch out how we might use this as a starting point of an iterative process to fit a functional model. Strictly speaking we should have modified our regression for Laboratory 1 by constraining the intercept term (α) to be 0, and the slope (β) to be 1. Having done this and fitted the first set of regression models, we can re-estimate the means (τs) for the nine samples from a weighted combination of the estimated value of τ for each

Table 4.21 Results of fitting measurement models to Mandel's pentose sugar data in *gllamm*.

(a) The structural measurement model for logged concentrations

Laboratory	Intercept, α (s.e.)	Slope, β (s.e.)	log SD of errors (s.e.)
1	0*	1*	−2.272 (0.136)
2	−0.088 (0.027)	1.022 (0.018)	−4.865 (0.198)
3	0.007 (0.028)	0.974 (0.019)	−3.100 (0.140)
4	−0.069 (0.030)	1.033 (0.020)	−3.019 (0.140)
5	−0.064 (0.034)	1.006 (0.023)	−2.470 (0.137)
6	−0.068 (0.027)	0.988 (0.018)	−3.490 (0.144)
7	−0.356 (0.062)	1.206 (0.042)	−1.567 (0.136)
μ_T	0.882 (0.088)		
σ_T^2	1.144 (0.133)		
logL	242.426		

* Constraint.

(b) The regression model for logged concentrations using the observed means of the 21 observations for each sample as an estimate of its true concentration

Laboratory	Intercept, α (s.e.)	Slope, β (s.e.)	log SD of errors (s.e.)
1	0.070 (0.030)	1.006 (0.021)	−2.115 (0.136)
2	−0.018 (0.015)	1.027 (0.010)	−2.820 (0.136)
3	0.075 (0.020)	0.979 (0.014)	−2.497 (0.136)
4	0.002 (0.015)	1.039 (0.010)	−2.801 (0.136)
5	0.005 (0.025)	1.012 (0.018)	−2.288 (0.136)
6	0.000 (0.014)	0.994 (0.010)	−2.846 (0.136)
7	−0.276 (0.047)	1.217 (0.033)	−1.668 (0.136)

(c) The structural measurement model for the raw (untransformed) concentrations, allowing for the error variance to increase with mean observed concentration

Laboratory	Intercept, α (s.e.)	Slope, β (s.e.)	log SD of errors (s.e.)
1	0*	1*	−1.383 (0.140)
2	−0.044 (0.043)	0.959 (0.016)	−2.941 (0.173)
3	0.043 (0.042)	0.950 (0.015)	−3.469 (0.253)
4	−0.067 (0.046)	1.002 (0.017)	−2.555 (0.156)
5	−0.016 (0.051)	0.952 (0.019)	−1.761 (0.141)
6	0.033 (0.041)	0.896 (0.015)	−2.717 (0.162)
7	−0.317 (0.056)	1.090 (0.021)	−1.801 (0.141)
γ**	0.506 (0.055)		
μ_T	4.604 (1.706)		
σ_T^2	26.099 (11.992)		
logL	127.481		

* Constraint.
** The slope of the line regressing the logarithm of the standard deviation of the measurement errors on the centred logarithm of the sample's mean sugar measurement.

measurement (that is, $(Y_{ijk} - \hat{\alpha}_i)/\hat{\beta}_j$), using the relevant estimated precision of the measurement ($\hat{\beta}_j^2/\hat{\sigma}_j^2$) as the weight:

$$\hat{\tau}_i = \sum_{j=1}^{7} \sum_{k=1}^{3} \frac{\hat{\beta}_j^2}{\sigma_j^2}\left[\frac{Y_{ijk} - \hat{\alpha}_i}{\hat{\beta}_j}\right] = \sum_{j=1}^{7} \sum_{k=1}^{3} \frac{\hat{\beta}_j}{\sigma_j^2}\left[Y_{ijk} - \hat{\alpha}_i\right] \qquad (4.25)$$

Here Y_{ijk} is the kth replicate for Laboratory j on sample i, τ_i is the true concentration for sample i, and $\hat{\alpha}_i$, $\hat{\beta}_j$ and $\hat{\sigma}_j^2$ are the current parameter estimates for method j (see Fuller, 1987, for example). Having re-estimated the τ_is we can then rerun the regressions, and so on, until convergence. The results are likely to be very close to those in Table 4.21(a). Readers interested in pursuing functional measurement error models are referred to Barnett (1970), Fuller (1987), Permutt *et al.* (1991), Kimura (1992), Bolfarine and Galea-Rojas (1995), and Gimenez and Bolfarine (2000). Kimura (1992) deals with the special case of no replications, and covers the situation where the observations might have a constant within-method error coefficient of variation in addition to the situation where the within-method error variance is assumed to be constant. Permutt *et al.* (1991) compare only two laboratories, but deal with the case of a varying number of replications. Much of this work is centred on the use of the EM-algorithm, and readers should also be aware of the more general application of EM algorithms to estimate parameters of measurement (factor analysis) models (see, for example, Dempster, Laird and Rubin, 1977; Rubin and Thayer, 1982 & 1983; Little and Rubin, 2002).

Finally, before leaving the pentose sugar data, we fit a measurement model to the raw (i.e. untransformed) data, allowing the variance of measurement errors to increase with the sugar concentration (to be precise, the logarithm of the standard deviation is assumed to be a linear function (with slope γ) of the centred logarithm of the observed mean (across all laboratories) of the sugar concentration). The model is equivalent to that fitted to the testicle volume data in Section 4.8. The results are shown in Table 4.21(c). Note that Laboratory 7 no longer appears to be quite such an outlier (although its intercept is still significantly different from zero).

We will now look at the dependence of the variance (or standard deviation) of the measurement errors on the amount of material being measured in a bit more detail. Consider Table 4.22(a), for example. This gives a series of measurements of the peak area using gas-chromatography/mass-spectroscopy for known concentrations of toluene (ranging from 4.6 picograms to 15 nanograms in $100\,\mu\text{L}$ of extract (these are very low concentrations – 1 picogram in $100\,\mu\text{L}$ corresponds to a concentration of 0.01 part per billion!). The data are from Rocke and Lorenzato (1995). Note that the variability of the measurements increases with toluene concentration (the standard deviation of the measurement errors being approximately proportional to the concentration of toluene) but that it does not shrink to zero at zero concentrations. Rocke and Lorenzato (1995) proposed a two-component model for the measurement errors for this situation. If P is the measured peak area obtained from a true toluene concentration of τ, then the proposed measurement model is:

$$P = \alpha + \beta\tau e^{\eta} + \varepsilon \qquad (4.26)$$

with intercept α and β as usual, but with two independently distributed measurement errors, η and ε, assumed to be normally distributed with zero means and variances σ_{η}^2 and σ_{ε}^2, respectively. The model is fitted iteratively using maximum likelihood or reweighted maximum marginal likelihood (see Rocke and Lorentzato, 1995, and Gibbons and Bhaumik, 2001, respectively, for technical details concerning these two methods). An alternative possibility for a measurement model has the following form:

$$\begin{aligned}
E(P|\tau) &= \alpha + \beta\tau \\
\text{Var}(P|\tau) &= \sigma^2(\alpha + \beta\tau)^{\theta}
\end{aligned} \qquad (4.27)$$

The second equation in (4.27) implies

$$\begin{aligned}
\log(\text{Var}(P|\tau)) &= 2\log(\sigma) + \gamma\log(\alpha + \beta\tau) \\
\log(\text{SD}(P|\tau)) &= \phi + \gamma\log(\alpha + \beta\tau)
\end{aligned} \qquad (4.28)$$

Table 4.22 Toluene measurements using gas-chromatography/mass-spectroscopy.

(a) The data

Toluene (pg)	Observed peak area			
4.6	29.80	16.85	16.68	19.52
23	44.60	48.13	42.27	34.78
116	207.70	222.40	172.88	207.51
580	894.67	821.30	773.40	936.93
3000	5350.65	4942.63	4315.79	3879.28
15000	20718.14	24781.61	22405.76	24863.91

Source: Rocke & Lorenzato (1995).

(b) Fitted values

Toluene (pg)	Peak mean		
	Observed	Model 1*	Model 2
4.6	20.71	18.52	19.73
23	42.45	46.56	47.85
116	202.62	188.29	189.98
580	856.58	895.43	899.11
3000	4622.09	4583.51	4597.60
15000	23192.35	22871.51	22937.20

Toluene (pg)	Peak standard deviation		
	Observed	Model 1*	Model 2
4.6	6.20	5.74	4.21
23	5.65	6.76	8.96
116	21.02	19.25	29.15
580	73.19	92.13	110.28
3000	652.98	475.65	445.98
15000	2005.02	2378.08	1766.01

* From the analysis of Rocke & Lorenzato (1995).

where $\phi = \log(\sigma)$ and $\gamma = \theta/2$. This is similar to that fitted to the testicle volumes (Section 4.8) and the pentose sugar concentrations (see above), except that the logarithm of the standard deviation of the measurement errors is proportional to $\log(\alpha + \beta\tau)$ rather than $\log(\tau)$. If α is approximately 0 and β approximately 1 then this distinction is fairly trivial, but in general it is probably preferable to use (4.27). This is a model that is thought to hold for a wide variety of immunoassays (see, for example, Rodbard, 1978; Davidian *et al.*, 1988; Carroll and Ruppert, 1988; and Carroll, 2003). Typically, θ takes values between 1 (Poisson-like variability) and 2 (gamma-like variability), corresponding to values of γ equal to 0.5 and 1, respectively.

Rocke and Lorenzato's estimates for the model described in Equation (4.26) – Model 1 – were $\hat{\alpha} = 11.51$, $\hat{\beta} = 1.524$, $\hat{\sigma}_\eta = 0.1032$ and $\hat{\sigma}_\varepsilon = 5.698$ (no standard errors were given). Fitting Equation (4.27) – Model 2 – using *gllamm* gives us the following estimates: $\hat{\alpha} = 12.697$ (s.e. 1.936), $\hat{\beta} = 1.528$ (s.e. 0.040), $\hat{\phi} = -1.117$ (s.e. 0.363) and $\hat{\gamma} = 0.856$ (s.e. 0.053). Fitting this model took place in two stages. The first round involved getting preliminary estimates of α and β by assuming that $\log(\text{SD errors}) = \phi + \gamma\log(\tau)$. These estimates were then used in a second stage using (4.28). A more refined estimate might have been obtained by iterating until convergence, but we didn't bother. The above estimate of

ϕ corresponds to $\hat{\sigma} = 0.327$. Table 4.22(b) gives the fitted values from the two competing models. There's little to choose between them.

Gibbons and Bhaumik (2001) have used the Rocke and Lorenzato two-component model to analyse the data from inter-laboratory precision studies. They assumed that the error model was common to all laboratories in the study (that is, the laboratories each had the same precision at each of the levels of material being measured) – they shared the same values of σ_η^2 and σ_ε^2. Instead of estimating each of the αs and βs for the different laboratories they assumed that the αs and βs were normally-distributed random variables with means μ_α and μ_β, respectively. The corresponding variances for α and β are σ_α^2 and σ_β^2, respectively. It is straightforward to fit a similar random effects regression model using the alternative model for the error variances (Equation 4.28) using software such as *gllamm*. Gibbons and Bhaumik (2001) illustrate their random effects regression modelling using the data from assays of copper shown in Table 4.23. We will not present the results of fitting

Table 4.23 Interlaboratory study of copper measurements.

Laboratory	Conc. (μ/L)	Replicate measurements (μ/L)					Mean	S.D.
1	0	3	2	−1	1	−1	0.800	1.789
1	2	3	3	5	2	2	3.000	1.225
1	10	14	10	11	12	13	12.000	1.581
1	50	54	51	52	54	38	49.800	6.723
1	200	205	206	208	211	195	205.000	6.042
2	0	2.1	0.3	2	1.3	2	1.540	0.764
2	2	8	1.8	0.7	4	3	3.500	2.805
2	10	10	12.4	10.6	12	11	11.200	0.990
2	50	53	54.6	50.3	50.1	50	51.600	2.089
2	200	188.6	210	210	214	200	204.520	10.293
3	0	0.8	−0.185	0.99	0.905	0.365	0.575	0.488
3	2	2.495	2.695	2.41	1.84	2.84	2.456	0.383
3	10	10.5	10.335	9.735	10.245	10.325	10.228	0.291
3	50	47.66	45.39	44.27	46.91	47.24	46.294	1.419
3	200	181.33	173.205	180.56	183.65	181.585	180.066	4.002
4	0	1.661	1.996	0	2.993	2.042	1.738	1.091
4	2	3.243	3.432	9.246	3.39	4.109	4.684	2.572
4	10	12.25	13.51	11.16	13.44	10.47	12.166	1.353
4	50	48.14	54.45	51.01	52.86	48.72	51.036	2.680
4	200	205.4	200.4	199.7	189.6	187.7	196.560	7.578
5	0	0.09	−2.51	7.27	7.14	0.28	2.454	4.475
5	2	0.86	2.68	−0.4	4.73	5.2	2.614	2.415
5	10	10.03	12.94	8.97	9.61	9.12	10.134	1.623
5	50	50.06	50.35	49.32	49.93	48.08	49.548	0.903
5	200	193.4	193.47	203.16	190.02	191.05	194.220	5.217
6	0	7.226	−1	0	10.244	−2.177	2.859	5.524
6	2	4.964	2	3	6.716	8.844	5.105	2.769
6	10	4.713	10	8	11.101	8.249	8.413	2.431
6	50	48.242	65	45	43	47	49.648	8.810
6	200	191.02	205	183	185	182	189.204	9.498
7	0	0.018	−3	0	−2	−2	−1.396	1.346
7	2	1.323	4.9	0	0	0	1.245	2.122
7	10	6	9.088	14.1	6	7	8.438	3.407
7	50	45.5	44	40	43	45.986	43.697	2.384
7	200	162	181	187	178.3	188.932	179.446	10.667

Source: Gibbons & Bhaumik (2001).

laboratory parameters as random effects but will stick to the fixed-effects approach. The reason for this decision will become clear as we describe the characteristics of the measurements. Looking at the column on the far right of Table 4.23 it is clear that for each laboratory the standard deviation of the measurements at zero copper concentration is not equal to 0 and the actual value of the standard deviation at zero dose may vary from one laboratory to another. It is also clear that the standard deviation of the measurements increases with increasing copper concentration, but that the extent of this increase might also vary from one laboratory to another (compare Laboratory 2 and Laboratory 5, for example). The assumption that the behaviour of the random measurement errors is the same for all of the laboratories appears to be untenable.

We start by fitting the, by now, standard measurement model in *gllamm* (except that in this example we will arbitrarily parameterize the model so that the αs are means rather than intercept terms). In terms of the model for the means of the measurements, each laboratory has its own characteristic values for both α and β. We use the two-component error model described in Equation (4.28), again with separate values for ϕ and γ for each of the seven laboratories. The first *gllamm* run, however, assumed that the logarithm of the standard deviation of the measurement errors was proportional to the logarithm of the true copper concentration (adding 0.1 to avoid trying to take logarithms of zero). The linear predictor was then extracted from the first run and used to revise the two-component error model for a second run (replacing any negative predictions for concentration of copper by 0.1 to avoid problems taking logarithms). We stopped at the results of the second run instead of iterating to convergence. The estimates from the second run are shown in Table 4.24(a).

The log-likelihood is -419.26 and that for the 1st run was -423.40. If we start with the results from the 1st run (including the linear predictors) and repeat the 2nd run with a variety of constraints we get the following. Constraining the αs to be equal for all seven laboratories we obtain a log-likelihood of -432.54. If, instead, we constrain the βs to be the same, we get a value of -446.94. Similarly, constraining either the ϕs to be equal or the γs to be equal gives log-likelihoods of -445.50 and -430.24, respectively. It appears that we should not be assuming that any of these constraints holds.

Returning to the data itself, one more point might be noticed. Within each of the laboratories the standard deviations for the measurements of 2 and $10\,\mu g/L$ of copper are often lower than those for zero concentration. Although it is difficult to see how this might be a valid reflection of reality (rather than chance) it does raise the possibility that the correct model for the standard deviations of the errors might be different (possibly more complex) than described by either (4.27) or (4.28). One possibility would be to use a quadratic function of the following form:

$$\log(SD(P\,|\,\tau)) = \phi + \gamma[\log(\alpha + \beta\tau)]^2 \tag{4.29}$$

or

$$\log(SD(P\,|\,\tau)) = \phi + \gamma_1\log(\alpha + \beta\tau) + \gamma_2[\log(\alpha + \beta\tau)]^2 \tag{4.30}$$

Re-running the second stage of our fitting procedure using (4.29) instead of (4.28) yields a log-likelihood of -415.43. It is difficult to tell the difference between (4.28) and (4.29) from these data. The parameter estimates for the revised model are given in Table 4.24(b). Using (4.30) yields a log-likelihood of -406.86. This appears to be an improvement but the results are perhaps difficult to interpret. The parameter estimates are given in Table 4.24(c). Notice how in every model Laboratory 5 stands out as being different, and in each

Table 4.24 Results of fitting the general measurement model to the copper assay data (2nd run).

(a) $\log(\mathrm{SD}(P|\tau)) = \phi + \gamma \log(\alpha + \beta\tau)$

Lab.	α (s.e.)	β (s.e.)	ϕ (s.e.)	γ (s.e.)
1	0.901 (0.440)	1.106 (0.014)	0.291 (0.223)	0.303 (0.068)
2	1.451 (0.384)	1.013 (0.015)	−0.037 (0.241)	0.378 (0.073)
3	0.658 (0.123)	0.901 (0.006)	−0.942 (0.198)	0.383 (0.059)
4	2.190 (0.421)	0.973 (0.012)	0.062 (0.254)	0.312 (0.077)
5	1.262 (0.738)	0.965 (0.008)	1.075 (0.210)	0.019 (0.060)
6	2.208 (1.107)	0.934 (0.020)	1.197 (0.267)	0.182 (0.082)
7	−1.118 (0.428)	0.903 (0.016)	0.725 (0.161)	0.242 (0.047)
$\log L$	−419.55			

(b) $\log(\mathrm{SD}(P|\tau)) = \phi + \gamma[\log(\alpha + \beta\tau)]^2$

Lab.	α (s.e.)	β (s.e.)	ϕ (s.e.)	γ (s.e.)
1	0.998 (0.438)	1.012 (0.019)	0.383 (0.221)	0.066 (0.017)
2	1.403 (0.411)	1.012 (0.016)	0.315 (0.189)	0.061 (0.012)
3	0.719 (0.120)	0.905 (0.008)	−0.968 (0.200)	0.086 (0.015)
4	2.263 (0.427)	0.973 (0.012)	0.368 (0.192)	0.051 (0.013)
5	1.265 (0.697)	0.965 (0.009)	1.012 (0.180)	0.010 (0.011)
6	2.094 (1.085)	0.935 (0.022)	1.358 (0.207)	0.032 (0.015)
7	−0.810 (0.548)	0.900 (0.017)	0.565 (0.195)	0.057 (0.013)
$\log L$	−415.43			

(c) $\log(\mathrm{SD}(P|\tau)) = \phi + \gamma_1[\log(\alpha + \beta\tau)] + \gamma_2[\log(\alpha + \beta\tau)]^2$

Lab.	α (s.e.)	β (s.e.)	ϕ (s.e.)	γ_1 (s.e.)	γ_2 (s.e.)
1	0.947 (0.443)	1.014 (0.017)	0.322 (0.237)	0.146 (0.268)	0.035 (0.058)
2	1.297 (0.453)	1.011 (0.016)	1.012 (0.695)	−0.664 (0.616)	0.163 (0.096)
3	0.759 (0.133)	0.907 (0.009)	−0.889 (0.242)	−0.203 (0.255)	0.126 (0.052)
4	2.361 (0.455)	0.972 (0.013)	0.875 (0.644)	−0.456 (0.538)	0.122 (0.085)
5	0.713 (0.461)	0.970 (0.009)	1.789 (0.364)	−1.164 (0.325)	0.208 (0.054)
6	2.010 (1.132)	0.936 (0.023)	1.476 (0.497)	−0.116 (0.440)	0.052 (0.076)
7	−1.001 (0.476)	0.902 (0.017)	0.591 (0.202)	0.159 (0.102)	0.025 (0.025)
$\log L$	−406.86				

case the differences can be seen as a reflection of what is happening to the observed standard deviations in Table 4.23. Our main conclusion, however, should not be that we have found the correct model, but that we should think very carefully about what a credible model might be. The final decision has to be influenced by both prior knowledge and the data at hand. We should not necessarily believe that errors will behave in a simple way. Equally important, we should be careful of over-fitting.

Before leaving this section it might be useful to mention one further point. The models have all been fitted to data with replicates. As in method comparison studies, we need to be sure that the measurement errors for these replicates are statistically independent. Jaech (1979) discusses two particular reasons why replicate data might provide misleading information concerning within-laboratory variability. The first is the temptation to 'clean' the data prior to the analysis. Obviously one wants to identify potential outliers in the data, but there is also a suggestion that by removing the discordant measurements the remaining replicates can be made to look more consistent than they really are. The second problem is caused by the fact that a laboratory might hold measurement conditions fixed while obtaining the replicate measurements. There are many sources of variability of measurements made within a laboratory, and under the assumption that the laboratory's quality will be judged largely on the basis of the closeness of the replicate measurements, the analyst may hold constant as many of the identified sources of variation as possible. This would lead to an underestimation of the within-laboratory variation normally associated with repeated measurements. In effect, it introduces correlated measurement errors.

4.12 Population heterogeneity

Table 4.25 presents a simple data set from a study of the severity of psychological distress in patients in primary care (Lewis *et al.*, 1992). It has also been analysed in some detail by Dunn (2000). Each patient is independently assessed and scored using three different methods: the General Health Questionnaire (GHQ: Goldberg, 1972), the Hospital Anxiety and Depression Scale (HADS: Zigmond & Snaith, 1983) and the revised version of Clinical Interview Schedule (CISR: Lewis *et al.*, 1992). Summary statistics are given in Table 4.26. The data are plotted in Figure 4.6. The data are positively skewed and the variability seems to be higher at higher levels of severity. This is reminiscent of the testicle volume and other data analysed earlier in this chapter, but in this case we will approach the analysis from a different perspective. We start by fitting the standard measurement model:

$$GHQ_i = \tau_i + \delta_i$$
$$HADS_i = \alpha_{HADS} + \beta_{HADS}\tau_i + \varepsilon_i \tag{4.31}$$
$$CISR_i = \alpha_{CISR} + \beta_{CISR}\tau_i + \upsilon_i$$

We could use the simple moment estimators described in Sections 4.2 and 4.4, or maximum likelihood assuming multivariate normality. The latter automatically gives us the associated standard errors. Table 4.27(a) shows the ML parameter estimates obtained using *Mplus*. The program set up is given in Appendix 4(a). (Similar results would have been obtained with other software such as *EQS* and *gllamm*, but we have switched to *Mplus* to illustrate the results of fitting the different measurement models to follow).

Clearly in this example the true distress scores are not normally distributed. The majority of the patients do not have marked psychological problems, but clearly there will be a significant minority who are likely to be unwell (suffering from depression, for example). We postulate that, in fact, our sample is a mixture of the two types (latent classes), each of which produces normally-distributed distress scores (with different means and variances). The resulting distribution is therefore an example of a finite mixture distribution (see, for example, Everitt and Hand, 1981; Titterington *et al.*, 1985; or McLacklan and Peel, 2000). We can fit such a finite mixture model using an ML EM algorithm in *Mplus* (Readers are referred to Muthén and Sheddon (1999) and Muthén and Muthén (1998–2001) for computational details). The parameters to be estimated are the mixing proportion (the proportion of patients who are unwell), the mean and variance of the true scores for the well patients, the mean and variance of the true scores for the unwell patients, and the familiar measurement model parameters from Equation (4.31). We start by assuming that the latter are identical in the two groups. The *Mplus* program set up is given in Appendix 4(b); the results are in Table 4.27(b). Note that the estimates of the parameters of the measurement error model

Table 4.25 Measurements of psychological distress (GHQ, HADS & CISR) scores.

Subject	GHQ	HADS	CISR	Subject	GHQ	HADS	CISR
1	17	17	25	53	12	13	32
2	8	4	9	54	11	13	21
3	12	14	9	55	12	14	17
4	11	5	6	56	17	13	24
5	30	18	30	57	6	3	3
6	9	7	11	58	6	18	7
7	16	4	14	59	8	3	4
8	6	4	3	60	10	11	10
9	11	7	18	61	17	12	12
10	8	8	15	62	6	1	2
11	10	4	4	63	33	35	28
12	16	16	16	64	17	11	7
13	6	1	1	65	20	16	25
14	14	13	19	66	4	4	9
15	13	14	12	67	11	14	9
16	8	5	6	68	5	4	5
17	29	30	49	69	13	12	17
18	16	18	9	70	14	29	32
19	12	7	14	71	5	4	22
20	11	14	12	72	9	6	7
21	8	6	9	73	8	10	4
22	8	6	1	74	18	20	17
23	7	6	2	75	22	27	33
24	23	15	28	76	18	14	16
25	7	5	3	77	13	14	24
26	20	6	11	78	15	24	21
27	8	17	7	79	14	12	16
28	15	21	37	80	19	16	16
29	19	20	26	81	12	9	2
30	13	14	13	82	11	2	2
31	6	9	12	83	8	6	20
32	16	21	26	84	7	2	6
33	16	5	4	85	6	3	2
34	15	16	23	86	12	14	10
35	7	3	3	87	11	11	14
36	8	7	4	88	14	9	13
37	8	6	5	89	15	12	11
38	16	12	10	90	9	11	14
39	27	20	22	91	11	10	18
40	1	1	0	92	9	7	5
41	32	28	36	93	7	1	1
42	9	11	20	94	10	5	9
43	6	12	13	95	11	11	12
44	11	6	9	96	8	7	8
45	13	20	16	97	9	1	4
46	9	12	6	98	17	12	32
47	7	5	5	99	12	17	15
48	11	17	23	100	15	10	13
49	13	9	12	101	9	9	11
50	19	23	35	102	5	0	3
51	12	10	18	103	17	11	23
52	7	8	3				

are practically unchanged. Interestingly the algorithm estimates that about 26% of the patients are in Group 2 (the unwell group). This is lower than would be expected using clinical criteria (estimates of around 50–55% are given for these data in Dunn, 2000, see page 44). Note that in terms of the goodness-of-fit criteria the results are not much of an

Table 4.26 Summary statistics for psychological distress (GHQ, HADS & CISR) scores ($N = 103$).

Variable	Mean	Std. Dev.	Min	Max
GHQ	12.311	5.967	1	33
HADS	11.068	7.064	0	35
CISR	13.709	9.860	0	49

Correlations

	GHQ	HADS	CISR
GHQ	1.000		
HADS	0.746	1.000	
CISR	0.715	0.774	1.000

Table 4.27 Measurement models for the psychological distress (GHQ, HADS & CISR) scores.

(a) A single group

Parameter	Estimate (s.e.)	Parameter	Estimate (s.e.)
α_{GHQ}	0*	β_{GHQ}	1*
α_{HADS}	-4.710 (1.586)	β_{HADS}	1.282 (0.122)
α_{CISR}	-7.383 (2.200)	β_{CISR}	1.713 (0.168)
σ^2_{GHQ}	10.962 (2.081)		
σ^2_{HADS}	9.508 (2.673)		
σ^2_{CISR}	24.951 (5.414)		
μ_T	12.311 (0.585)	σ^2_T	24.300 (4.880)
logL	-963.396		
AIC	1944.792		
BIC	1968.505		

(b) A two-group finite mixture model (equal error variances across groups)

Parameter	Estimate (s.e.)	Parameter	Estimate (s.e.)
α_{GHQ}	0*	β_{GHQ}	1*
α_{HADS}	-4.332 (1.641)	β_{HADS}	1.251 (0.106)
α_{CISR}	-7.246 (2.159)	β_{CISR}	1.702 (0.130)
σ^2_{GHQ}	10.415 (2.290)		
σ^2_{HADS}	10.540 (2.640)		
σ^2_{CISR}	24.289 (4.895)		
GROUP 1			
μ_T	10.358 (1.308)	σ^2_T	8.571 (7.531)
GROUP 2			
μ_T	17.976 (17.114)	σ^2_T	28.913 (80.163)
Proportion in GROUP 2		0.26 (logit(P) = -1.065 (s.e. 3.502))	
logL	-955.562		
AIC	1935.124		
BIC	1928.835		

(continued)

Table 4.27 (*continued*)

(c) A two-group finite mixture model (different error variances in the two groups)

Parameter	Estimate (s.e.)	Parameter	Estimate (s.e.)
α_{GHQ}	0*	β_{GHQ}	1*
α_{HADS}	-4.858 (1.006)	β_{HADS}	1.294 (0.101)
α_{CISR}	-8.322 (1.202)	β_{CISR}	1.773 (0.122)
GROUP 1			
σ^2_{GHQ}	-0.122 (0.287)		
σ^2_{HADS}	3.667 (1.430)		
σ^2_{CISR}	4.570 (2.105)		
μ_T	7.457 (0.321)	σ^2_T	1.546 (0.805)
GROUP 2			
σ^2_{GHQ}	14.501 (3.712)		
σ^2_{HADS}	11.837 (4.226)		
σ^2_{CISR}	29.634 (7.487)		
μ_T	13.812 (0.789)	σ^2_T	20.767 (4.189)
Proportion in GROUP 2		0.75 (logit(P) = 1.202 (s.e. 0.284))	
$\log L$	-935.906		
AIC	1901.812		
BIC	1941.333		

(d) A two-group finite mixture model (different error variances in the two groups, but error variances for GHQ and HADS constrained to be equal)

Parameter	Estimate (s.e.)	Parameter	Estimate (s.e.)
α_{GHQ}	0*	β_{GHQ}	1*
α_{HADS}	-5.800 (1.183)	β_{HADS}	1.349 (0.102)
α_{CISR}	-8.814 (1.560)	β_{CISR}	1.805 (0.136)
GROUP 1			
σ^2_{GHQ}	1.800 (0.727)		
σ^2_{HADS}	1.800 (0.727)		
σ^2_{CISR}	3.003 (1.811)		
μ_T	7.533 (0.461)	σ^2_T	1.475 (0.974)
GROUP 2			
σ^2_{GHQ}	13.654 (2.076)		
σ^2_{HADS}	13.654 (2.076)		
σ^2_{CISR}	31.000 (8.165)		
μ_T	14.129 (0.878)	σ^2_T	18.473 (4.091)
Proportion in GROUP 2		0.75 (logit(P) = 1.073 (s.e. 0.313))	
$\log L$	-938.739		
AIC	1903.478		
BIC	1937.729		

improvement on the initial one-group model. Note that the mean and variance of the true scores for Group 2 are very poorly estimated.

We now fit a further model in which we allow the error variances for the GHQ, HADS and CISR to differ in the two latent classes. The *MPlus* set up is given in Appendix 4(c);

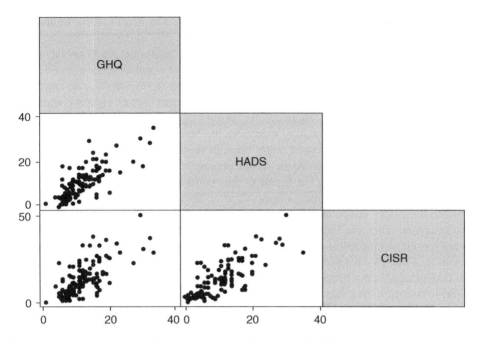

Figure 4.6 Scatterplot matrix for three measures of psychological distress.

the results are in Table 4.27(c). There has been a fairly dramatic improvement in fit, but the proportion of unwell patients is now estimated to be about 75%. The mean and variance of the true scores for Group 2 are now much better estimated. But we have an embarrassing negative variance estimate for the errors for the GHQ in Group 1. Note that the Group 2 error variances for GHQ and HADS are very similar. We solve the negative error variance problem by simply constraining these error variances to be equal in Group 1 and also equal to each other in Group 2. The resulting log-likelihood is -938.739 (twice the difference being 5.666 with 2 d.f.), indicating that this is a fairly reasonable solution. The parameter estimates are given in Table 4.27(d). Perhaps a slightly more sensible constraint would have been to constrain the precisions to be equal (i.e. $\sigma^2_{GHQ} = \sigma^2_{HADS}/\beta^2_{HADS}$) but the outcome would have been essentially the same.

In the above discussion we have considered that it is a distinct possibility that there are different groups of patients (based on prior knowledge concerning what we are measuring) and the behaviour of the measurement model for these groups is of substantive interest in its own right. Readers interested in pursuing this area are referred to Yung (1997). We may, however, have no expectation that the population under study is heterogeneous in this way but simply use finite mixture distributions to make the assumptions concerning the distribution of the true scores more realistic. Carroll, Roeder and Wasserman (1999) and Richardson, Leblond, Jaussant and Green (2002) have proposed the use of finite mixture distributions for this purpose. If we have no missing data the results of simply fitting the measurement model to a single group (i.e. assuming normality of the true scores) is not likely to lead to problems (hence the apparent lack of concern about this in earlier examples), but if there are missing data then this might not be the case (see Carroll, Freedman and Pee, 1997). We will very briefly return to this topic at the end of Section 4.13. Other methods of coping with non-normal data are discussed in detail by Mooijaart (1985), Muthén and Kaplan (1992), and by Yuan and Bentler (1998 and 2000).

Consider a hypothetical set of measurements made in an analytical chemistry laboratory. The first two, A and B, do not involve the destruction of the material under test and can be made on all specimens of material. The second two, C and D, both involve destruction of the specimen. A particular specimen can yield a measurement C or D, but not both. A simple design for a precision study might involve taking measurements on all specimens using A and B and then randomly allocating specimens to obtain either measurement C or measurement D. An analogous situation might arise in the case of a psychiatric assessment where, in this situation, A and B are the results obtained from two fairly dissimilar psychiatric screening questionnaires, but C and D are results obtained through the use of two closely related and time-consuming diagnostic interviews. Here for reasons of cost as well as potential memory effects it might be thought impossible to expose subjects to both of these interviews (but see Lewis *et al.*, 1992). How do we analyse data such as these? Having got our two groups of specimens (or subjects) – those with A, B and C, and the others with A, B and D – we need to simultaneously fit our measurement models to the data from both groups. In this particular example we are justified (from the randomization in the design) to constrain the mean and the variance of the true scores to be the same in the two groups. We also constrain the parameters associated with A and B to be common to both groups. This then allows us to estimate the parameters for both C and D and to test which, if any, might be the same (are the intercepts or slopes, for example similar for C and D, or are their precisions the same?). If the fitting criterion for group g (here $g = 1$ or 2, but in general there might be more than two groups) is $F_{ML(g)}$ (that is, F_{ML} for group g as defined by Equation (4.12)) then the overall fitting criterion is given by

$$F_{ML} = \sum_{g=1}^{M} \frac{N_g}{N} F_{ML(g)} \tag{4.32}$$

where N_g is the number of specimens (subjects) in group g, M is the number of groups, and N is the sum of the N_gs.

Table 4.28 contains summary statistics from the results of a computer-simulated BIBD reliability study with 50 subjects per group. Each group provides three measurements out of a possible four (A, B, C and D). Details of the simulation are given in Dunn (1989, Table 5.21). We can fit a measurement model to these four groups simultaneously. Each group has a measurement model for three of the measurements. The mean and variance of the true scores are constrained to be equal across all four groups. The parameters involving A in groups 1, 2 and 3 are all constrained to be equal across these three groups. Similar constraints are used for each of the other three measures, B, C and D. The justification for all of these constraints is again randomization. The imposition of all of these constraints in *EQS* and similar software is a bit tedious, but quite straightforward. The parameter estimates are given in Table 4.29.

In general our data from multiple groups may not be obtained using such a tight design as those described above. We might, for example, wish to compare measurement models for men and women, or for two or more species of bacteria, or analytical measurements of chemicals embedded within different media, and so on. Depending on the context we may or may not be justified in the introduction of cross-group constraints. And, of course, we can always empirically test whether such constraints can be introduced without destroying the fit of the model. The modelling strategies to be adopted should be fairly clear.

Table 4.28 Results of a simulated BIBD reliability study with 50 subjects per group.

Group 1

| | Covariance matrix | | | Mean |
	A	B	C	
A	13.559			2.311
B	19.431	34.733		3.625
C	31.566	50.173	89.000	4.227

Group 2

| | Covariance matrix | | | Mean |
	A	C	D	
A	11.550			1.949
C	33.347	113.719		4.112
D	11.322	36.007	13.719	1.719

Group 3

| | Covariance matrix | | | Mean |
	A	B	D	
A	10.659			2.132
B	18.439	38.804		2.954
D	9.509	17.952	10.456	1.830

Group 4

| | Covariance matrix | | | Mean |
	B	C	D	
B	24.016			2.355
C	29.342	56.632		2.763
D	9.343	13.403	5.088	1.843

Table 4.29 Results of fitting the standard measurement model to the BIBD data.

Parameter	Estimate (s.e.)
α_A	0*
α_B	−0.780 (0.299)
α_C	−2.174 (0.474)
α_D	−0.099 (0.161)
β_A	1*
β_B	1.817 (0.084)
β_C	2.864 (0.132)
β_D	0.963 (0.045)
σ_A^2	1.163 (0.216)
σ_B^2	3.964 (0.724)
σ_C^2	10.161 (1.828)
σ_D^2	1.222 (0.214)
σ_τ^2	9.518 (1.134)
μ_τ	2.055 (0.239)

* Constraint.
Chi-square = 31.488 with 24 d.f.

4.13 Data missing by design

The focus of the present section will be on the use of two-phase designs for method comparison studies. The illustrative examples will be based on work in nutritional epidemiology in which the investigators carry out what they term 'calibration' or 'validation' studies as part of a large analytical survey to explore the relationships between certain aspects of diet and risk of subsequent illness. Their primary aim is to be able to correct for measurement errors in the estimation of the influence of diet on risk of disease (see, for example, Freedman, Carroll and Wax, 1991; Plummer and Clayton 1993a, 1993b; Kaaks *et al.*, 1994, Carroll, Freedman and Pee, 1997; Wong, Day, Bashir and Duffy, 1999 and Wong, Day and Wareham, 1999). Two-phase designs also have a long history of use in psychiatric epidemiology but, in this application, the investigators are more concerned with errors in the assessment of illness, and measurement of illness severity, rather than in exposure to potential risk factors (see, for example, Dunn, Everitt and Pickles, 1993 and Dunn, 2000).

Nutritional epidemiologists are typically concerned with the measurement of an individual's usual or 'average' diet. But in Western societies, in particular, there are large day-to-day variations in what people eat, and even if we were to measure exactly what a person ate on a given day it would still be a very fallible estimate of that person's usual intake (with a reliability coefficient at most 0.5 – see Carroll *et al.*, 1997, for example). From its very nature, usual dietary intake cannot be measured with high (or even moderate) precision. One of the most convenient and inexpensive methods of measurement involves the use of a food frequency questionnaire (FFQ). FFQs contain lists of foods commonly eaten by the people being studied (meats, fruits and vegetables, cereals, and so on) and the person completing the questionnaire is asked to work through the list and indicate the frequency in which they have eaten the different foods over, say, the last 6 or 12 months. Sometimes the respondents are also asked to indicate the size of each food portion consumed. There are many potential sources of error arising from the use of a FFQ: over- or under-reporting, memory problems, lack of awareness of the sizes of the portions consumed, and so on.

So, if we were to try to evaluate both the systematic biases and the random components of the measurement errors associated with the FFQ, what options do we have? Some information can be gained by repeating the use of the FFQ, but this would tell us nothing about systematic under- or over-reporting or about equation errors (subject-specific biases). We need to use different methods of measurement. One common alternative is the use of a 24-hour food recall (FR), which is obtained through the use of a structured interview. This method is much more time-consuming and expensive than using the FFQ but it is thought to be subject to less bias. Although it might provide a precise estimate of a given day's intake, it will still, however, be a fallible measure of usual intake for the reasons given above. Now, to obtain an identified measurement model we need to replicate the FR measurement or find a good instrumental variable (a biomarker, for example – see Wong *et al.*, 1999a, b). Repeating the FR measure will again be time consuming and expensive, and using a biomarker, in particular, could be very expensive (depending on the choice of laboratory assay technique for a given biomarker). For this reason investigators will not make all of the measurements on all of the participants. They will choose to have data missing by design. This is the rationale for a two-phase selection procedure.

Phase 1 involves giving everyone in the main epidemiological survey an FFQ to complete. Typically the 1st phase sample comprises several thousand participants. The 2nd phase involves the selection of a sub-sample of participants from those in phase 1, the size of the 2nd phase sample being considerably smaller than that for phase 1. The phase 2 participants are then asked to complete, say, their 1st 24-hour food recall (FR1) interview (after sufficient time has elapsed after the completion of the FFQ to reduce contamination to a minimum). After a further month or so, the 2nd phase participants are asked to complete another food recall interview (FR2). A biomarker measurement might be taken instead of, or in addition to, FR2. It is also possible that the number of food recall measurements might be more than two. This raises the question concerning whether it is more efficient to use a few FR measurements on many phase 2 participants, or many FR measurements on far fewer participants (see Carroll *et al.*, 1997). Of course, it is also possible to add further FFQ measurements and this will allow us to assess whether there are equation errors associated with the FFQ. Here, however, we only consider the simple design in which we have FFQ measurements on all phase 1 participants and two independent FR measurements on those selected for phase 2. The question we then ask is 'How should we analyze the data?'

We are faced with a missing data problem, but because we have determined the pattern of missing values through the design, we know that the missing observations are either missing completely at random (MCAR) or missing at random (MAR), depending on exactly how we have selected the 2nd phase participants. For MCAR we simply select a random sub-set of the phase 1 participants without reference to any of their characteristics (including the FFQ measurement). For MAR we have typically stratified the 1st phase participants on the basis of the FFQ measurements (and possibly other covariates) and selected sub-samples from each of these strata (usually varying the sampling fractions across strata to ensure that the more interesting subjects are well represented in the 2nd phase sample). We have already seen in Section 4.12 that we can use multiple group analyses to analyse data arising from designs in which the missing data are MCAR (the BIBD, for example) and this method can also be used in the present situation. Even when the 2nd phase data are MAR, rather than MCAR, we proceed by constraining all of the relevant parameters to be the same in the two groups (Group 1 being the 2nd phase sample with measurements on FFQ, FR1 and FR2; Group 2 being those 1st phase participants who provide only that from the FFQ). The case of data which are MCAR was first considered by Werts *et al.* (1979). Muthén, Kaplan and Hollis (1987) extended the method for data with an ignorable missing data mechanism (MAR as well as MCAR). A detailed illustration of the use of this approach for a simple measurement error model is provided by Dunn *et al.* (1993: Chapter 8). Here we use a more direct method of maximum likelihood estimation as implemented in *MPlus*, which is based on the EM algorithm as discussed in Little and Rubin (2002 – see Chapter 11). The fitting function described in Equation (4.12) is modified as follows:

$$F_{ML} = \sum_{i=1}^{N}\left[\log\left|\hat{\boldsymbol{\Sigma}}_i\right| + \left(\boldsymbol{v}_i - \hat{\mu}_i\right)'\hat{\boldsymbol{\Sigma}}_i^{-1}\left(\boldsymbol{v}_i - \hat{\mu}_i\right) - c\right] \tag{4.33}$$

where $\boldsymbol{v}_i$ is the vector of observed measurements for the ith phase 1 participant (N being the total phase 1 sample size) and $\hat{\mu}_i$ is the corresponding vector of values predicted by the fitted model. The dimensions of $\boldsymbol{v}_i$, $\hat{\mu}_i$ and $\hat{\boldsymbol{\Sigma}}_i$ depend on the number of non-missing observations for the ith subject (in this example there will be only two patterns: one for three non-missing observations in phase 2 and the other for the one observation for the phase 1 participants who were not selected for phase 2). The last term in Equation 4.33 (i.e. c) is an offset determined by fitting a completely unstructured model to the same data (for technical details, see Appendix 6 of Muthén and Muthén, 1998–2001). For the more general case where we have missing data within each of multiple groups (men and women, different age groups, and so on) Equation (4.33) can be extended as follows:

$$F_{ML} = \sum_{g=1}^{G} \sum_{i=1}^{n_g} \left[\log \left| \hat{\mathbf{\Sigma}}_{gi} \right| + \left(\mathbf{v}_{gi} - \hat{\mu}_{gi} \right)' \hat{\mathbf{\Sigma}}_{gi}^{-1} \left(\mathbf{v}_{gi} - \hat{\mu}_{gi} \right) - c_g \right] \qquad (4.34)$$

The notation should be clear from the context (g indicating group membership, with a total of G groups). Again, see Muthén & Muthén (1998–2001) for technical details. Here we concentrate on illustrating the implications of the design for the analysis of the data and interpretation of the results.

For illustation we will use some data generated by *Monte Carlo* simulation. We have 1000 phase 1 subjects, each with an FFQ measurement. We then select a sub-sample of these participants using Bernoulli sampling with a fixed selection probability of 0.2 (i.e. the resulting data are MCAR). Each of the 203 phase 2 subjects then provide measurements for both FR1 and FR2. The measurement model that holds for these data is the following:

$$FR1_i = \tau_i + \delta_i$$
$$FR2_i = \tau_i + \varepsilon_i \qquad (4.35)$$
$$FFQ_i = \alpha_{FFQ} + \beta_{FFQ}\tau_i + \nu_i$$

with $E(\tau) = \mu_\tau = 40$, $\mathrm{var}(\tau) = \sigma_\tau^2 = 25$, $\mathrm{var}(\delta_i) = \mathrm{var}(\varepsilon_i) = \sigma_{FR}^2 = 25$, $\mathrm{var}(\nu_i) = \sigma_{FFQ}^2 = 40$, $\alpha_{FFQ} = 5$, and $\beta_{FFQ} = 0.80$. These parameter values have been selected to more-or-less correspond to those given in Carroll *et al.*, 1997. Summary statistics for the resulting data are given in Table 4.30 (which also includes summary statistics for the equivalent data set without missing values). Note the very modest correlations. We fit the model in (4.35) three times: first to the full data set without missing values, then to the participants with data on all three measurements (a naïve complete case analysis), and finally to all of the available data using (4.33) as the fitting criterion. *MPlus* is used in all cases (see Appendix 4(d)). The results are given in Table 4.31. We see that in the case where the MCAR assumption holds, the complete case analysis is not leading us astray. The more sophisticated approach to fitting the model to all available data is simply adding a little extra efficiency.

Now we change the design so that missing data are MAR but not MCAR. We keep the same 1st phase sample but for the 2nd phase sample we use Bernoulli sampling with selection probability of 0.4 when FFQ is greater than 37, but 0 otherwise. That is, we include none of the 1st phase participants with a low FFQ in our 2nd phase sample. The summary statistics obtained from data generated by the second design are given in Table 4.32 (again we include the same summary statistics for the data without missing values for comparison). Note the expected attenuation of both the correlations and standard deviations within the phase 2 sample. The results of fitting model (4.35) to (a) the complete cases, and (b) all available data assuming MAR, are shown in Table 4.33. The key point to note is that the complete case approach (analysis of phase 2 data only) is now performing very poorly. The estimate for α is far higher than expected and that for β is lower. The measurement error variance estimates are also rather low. Approach (b), however, is performing quite well.

We now try a third two-phase design. Keeping the same phase 1 data, we simply select all participants with an FFQ measurement of less than 27 (the bottom 10%) or more than 47 (the top 10%). The middle 80% are not represented at all in the 2nd phase. Summary

Table 4.30 Simulated two-phase data (MCAR).

(a) Summary statistics for full data set (no missing data: $N = 1000$)

Variable	Mean	Standard deviation
FFQ	37.209	7.445
FR1	39.998	7.154
FR2	40.131	7.206

Correlations

	FFQ	FR1	FR2
FFQ	1		
FR1	0.3413	1	
FR2	0.3442	0.4911	1

(b) Summary statistics for Group 1 (Phase 2 participants: $N = 203$)

Variable	Mean	Standard deviation
FFQ	37.340	7.477
FR1	40.317	7.243
FR2	40.445	7.600

Correlations

	FFQ	FR1	FR2
FFQ	1		
FR1	0.3445	1	
FR2	0.3996	0.4644	1

(c) Summary statistics for Group 2 (Phase 1 participants only: $N = 797$)

Variable	Mean	Standard deviation
FFQ	37.176	7.441

Table 4.31 Analysis of two-phase data from Table 4.30 (MCAR).

Parameter	Estimate (standard error)		
	Full data set (no missing values)	Complete cases (Phase 2 only)	All available data (assuming MAR)
α_{FFQ}	8.209 (2.335)	4.662 (5.454)	4.676 (5.041)
β_{FFQ}	0.724 (0.058)	0.809 (0.135)	0.807 (0.125)
μ_τ	40.064 (0.196)	40.381 (0.445)	40.333 (0.410)
σ_τ^2	25.287 (1.815)	25.430 (4.243)	25.395 (4.186)
σ_{FR}^2	26.222 (1.173)	29.410 (2.919)	29.410 (2.919)
σ_{FFQ}^2	42.125 (2.313)	38.978 (4.918)	38.852 (4.262)

statistics are given in Table 4.34. Not that the magnitude of the correlations and standard deviations has been increased by the phase 2 selection process. Although this is an extreme form of phase 2 selection it corresponds to a familiar approach to optimizing the regression for calibration problems in which the explanatory variable is not subject to measurement error. The results of the two model fitting approaches are shown in Table 4.35. Now we see that the complete case analysis leads to over estimation of β and the three variances, and

Table 4.32 Simulated two-phase data (MAR – version 1).

(a) Summary statistics for full data set (no missing data: $N = 1000$)

Variable	Mean	Standard deviation
FFQ	37.209	7.445
FR1	39.998	7.154
FR2	40.131	7.206

Correlations

	FFQ	FR1	FR2
FFQ	1		
FR1	0.3413	1	
FR2	0.3442	0.4911	1

(b) Summary statistics for Group 1 (Phase 2 participants: $N = 212$)

Variable	Mean	Standard deviation
FFQ	43.013	4.376
FR1	42.031	6.939
FR2	41.834	7.260

Correlations

	FFQ	FR1	FR2
FFQ	1		
FR1	0.1992	1	
FR2	0.2210	0.4306	1

(c) Summary statistics for Group 2 (Phase 1 participants only: $N = 788$)

Variable	Mean	Standard deviation
FFQ	35.648	7.329

Table 4.33 Analysis of two-phase data from Table 4.32 (MAR – version 1).

	Estimate (standard error)		
Parameter	Full data set (no missing values)	Complete cases (Phase 2 only)	All available data (assuming MAR)
α_{FFQ}	8.209 (2.335)	30.374 (3.584)	7.956 (4.752)
β_{FFQ}	0.724 (0.058)	0.301 (0.085)	0.732 (0.119)
μ_τ	40.064 (0.196)	41.933 (0.411)	39.952 (0.407)
σ_τ^2	25.287 (1.815)	21.578 (3.753)	25.806 (4.100)
σ_{FR}^2	26.222 (1.173)	17.100 (1.792)	28.623 (2.780)
σ_{FFQ}^2	42.125 (2.313)	28.623 (2.780)	41.539 (4.023)

underestimates α (in general the estimates of α and β are negatively correlated). Again, the approach based on all available data is performing well, as expected.

Although the message should be very clear by now, we will reinforce it by reference to one final data set. Using the same phase 1 sample as before, we generate FR1 for all phase 1

Table 4.34 Simulated two-phase data (MAR – version 2).

(a) Summary statistics for full data set (no missing data: $N = 1000$)

Variable	Mean	Standard deviation
FFQ	37.209	7.445
FR1	39.998	7.154
FR2	40.131	7.206

Correlations

	FFQ	FR1	FR2
FFQ	1		
FR1	0.3413	1	
FR2	0.3442	0.4911	1

(b) Summary Statistics for Group 1 (Phase 2 participants: $N = 190$)

Variable	Mean	Standard deviation
FFQ	36.352	13.432
FR1	39.590	7.944
FR2	39.513	7.810

Correlations

	FFQ	FR1	FR2
FFQ	1		
FR1	0.5273	1	
FR2	0.5023	0.6237	1

(c) Summary Statistics for Group 2 (Phase 1 participants only: $N = 810$)

Variable	Mean	Standard deviation
FFQ	37.410	5.107

Table 4.35 Analysis of two-phase data from Table 4.34 (MAR – version 2).

Parameter	Estimate (standard error)		
	Full data set (no missing values)	Complete cases (Phase 2 only)	All available data (assuming MAR)
α_{FFQ}	8.209 (2.335)	−19.336 (6.345)	12.712 (4.486)
β_{FFQ}	0.724 (0.058)	1.408 (0.159)	0.615 (0.112)
μ_τ	40.064 (0.196)	39.551 (0.514)	39.810 (0.427)
σ_τ^2	25.287 (1.815)	38.492 (5.277)	27.174 (4.128)
σ_{FR}^2	26.222 (1.173)	23.234 (2.384)	23.235 (2.384)
σ_{FFQ}^2	42.125 (2.313)	103.161 (13.161)	45.085 (3.807)

participants (that is, everyone now has FFQ and FR1). We now select the bottom 10% and top 10% of the FFQ scorers for phase 2 and generate an instrumental variable (IV) according to

$$IV_i = \alpha_{IV} + \beta_{IV}\tau_i + \omega_i \tag{4.36}$$

Table 4.36 Simulated two-phase data (MAR – version 3).

(a) Summary statistics for full data set (no missing data: $N = 1000$)

Variable	Mean	Standard deviation
FFQ	37.209	7.445
FR1	39.998	7.154
FR2	40.131	7.206

Correlations

	FFQ	FR1	FR2
FFQ	1		
FR1	0.3413	1	
FR2	0.3442	0.4911	1

(b) Summary Statistics for Group 1 (Phase 2 participants: $N = 190$)

Variable	Mean	Standard deviation
FFQ	36.352	13.432
FR1	39.590	7.944
FR2	81.224	12.545

Correlations

	FFQ	FR1	FR2
FFQ	1		
FR1	0.5273	1	
FR2	0.6244	0.7618	1

(c) Summary Statistics for Group 2 (Phase 1 participants only: $N = 810$)

Variable	Mean	Standard deviation
FFQ	37.410	5.107
FR1	40.093	6.958

Correlations

	FFQ	FR1
FFQ	1	
FR1	0.2589	1

in which $\alpha_{IV} = 10$, $\beta_{IV} = 1.8$ and $\text{var}(\omega) = \sigma_{IV}^2 = 25$. Summary statistics are given in Table 4.36. We fitted the appropriate measurement error model to (a) a data set without any missing data – that is, with the IV measurements available for all phase 2 subjects, (b) only the phase 2 subjects (the complete case analysis) and (c) all available cases assuming MAR – see Table 4.37. These results need no further comment!

In some of the above examples we have used extreme sampling fractions for selection into phase 2 (i.e. they have been either 0 or 1 for each of the phase 1 strata). In practice such

Table 4.37 Analysis of two-phase data from Table 4.36 (MAR – version 3).

Parameter	Estimate (standard error)		
	Full data set (no missing values)	Complete cases (Phase 2 only)	All available data (assuming MAR)
α_{FFQ}	9.636 (2.011)	−18.516 (5.794)	12.489 (3.202)
α_{IV}	7.588 (5.306)	7.192 (6.688)	8.875 (9.347)
β_{FFQ}	0.689 (0.050)	1.386 (0.145)	0.618 (0.080)
β_{IV}	1.862 (0.132)	1.870 (0.168)	1.825 (0.233)
μ_τ	39.998 (0.226)	39.590 (0.575)	39.998 (0.226)
σ_τ^2	26.344 (2.617)	40.387 (6.648)	29.384 (4.144)
σ_{TR}^2	24.789 (2.019)	22.385 (3.642)	21.748 (3.720)
σ_{FFQ}^2	42.854 (2.078)	101.897 (11.7800)	44.150 (2.404)
σ_{IV}^2	18.453 (5.910)	15.335 (10.010)	17.393 (11.579)

sampling fractions are unlikely to be used (at least in the large nutritional epidemiology surveys which we have used as a model). Carroll *et al.*, 1997, for example, discuss stratifying into five groups on the basis of FFQ: 0–25, 25–32.5, 32.5–40, 40–47.5 and >47.5. They use 2nd-phase sampling fractions of 0.20, 0.15, 0.10, 0.15 and 0.40 for each of these strata, respectively. Data from such stratified two phase calibration studies can be analysed by maximum likelihood (as above), but it is also possible to use moments-based estimators, after correcting for biases arising from the missing data mechanism through the use of inverse probability weights (Horvitz-Thompson estimators). Inverse probability weighting will be discussed briefly in Chapter 5 (see also Dunn, 2000) but we refer readers to the paper by Carroll *et al.* for details of their application in the present context. Note that they cannot be used if any of the phase 1 stratum sampling probabilities are 0.

Finally, we briefly mention the issue of multivariate normality. The methods described in this section are dependent on the normality assumption and are likely to produce biased results if this assumption does not hold (see Carroll *et al.*, 1997). The use of finite mixture distributions as suggested by Carroll, Roeder and Wasserman (1999) and discussed above (Section 4.12) can help here, and it is possible to use ML EM-based software such as *MPlus* to fit measurement error models incorporating both finite mixtures and data missing by design.

4.14 Multitrait-multimethod designs

So far in this book we have mainly been concerned with reliability and precision. Here we will try to incorporate ideas concerning validity. It will inevitably involve the use of some of the terminology of the factor analysis tradition, but we hope that this will not be too distracting for the reader. Much of the following discussion of the measurement of child psychiatric problems is based on that in Dunn (2000). However, we also illustrate the use of the methodology for the analysis of measurements other than in the social sciences, and illustrate the ideas using blood pressure measurements.

In a very influential social science paper, Campbell and Fiske (1959) produce the following statement:

> Reliability is the agreement between two efforts to measure the same trait through maximally similar methods. Validity is represented in the agreement between two methods to measure the same trait through maximally different methods.

They went on to suggest that reliability and validity should be jointly assessed through the use of a multitrait-multimethod (MTMM) approach. Here two or more traits or character-istics in a sample of participants are assessed using two or more measurement techniques (ideally based on completely different principles). In child psychiatry, for example, one might be interested in measuring hyperactivity and the extent of conduct disorder using information provided by multiple informants (peer, teacher or parent, for example). The consistency of the teachers' reports gives one a handle on teachers' reliability; the correla-tion or association between the views of teachers and parents tells us about validity.

Table 4.38(a) is an example of correlation matrix derived from the results of a multitrait-multimethod study of childhood anxiety and depression (Cole *et al.*, 1997). Table 4.38(b) is the correlation matrix from a similar study of childhood depression and aggression (Messer and Gross, 1994). The correlations from Cole *et al.* are those obtained after deleting over-lapping items from the assessment rating scales. The striking thing about these two matri-ces is their similarity. Almost equally striking is the fact that the within-method, cross-trait correlations (those in bold) are relatively high compared with the cross-method, within-trait correlations (those underlined). High cross-method, within trait correlations are

Table 4.38 Examples of multitrait-multimethod correlation matrices.

(a) Childhood depression and anxiety (from Cole *et al.*, 1997)

	V1	V2	V3	V4	V5	V6	V7	V8
Depression								
V1 – Self	1.00							
V2 – Peers	0.21	1.00						
V3 – Teacher	0.19	0.32	1.00					
V4 – Parent	0.21	0.32	0.27	1.00				
Anxiety								
V5 – Self	**0.66**	0.17	0.10	0.16	1.00			
V6 – Peers	0.18	**0.72**	0.26	0.22	0.10	1.00		
V7 – Teacher	0.15	0.25	**0.76**	0.25	0.06	0.27	1.00	
V8 – Parent	0.24	0.33	0.22	**0.72**	0.22	0.25	0.20	1.00
N = 280								
Mean	5.98	0.45	17.36	4.23	15.31	0.53	16.37	10.62
S.D.	7.21	0.43	6.02	4.56	9.44	0.39	5.51	6.24

(b) Childhood depression and aggression (from Messer & Gross, 1994)

	V1	V2	V3	V4	V5	V6
Depression						
V1 – Self	1.00					
V2 – Peers	0.24	1.00				
V3 – Teacher	0.19	0.25	1.00			
Aggression						
V4 – Self	**0.47**	0.13	0.07	1.00		
V5 – Peers	0.11	**0.49**	0.16	0.24	1.00	
V6 – Teacher	0.16	0.33	**0.47**	0.28	0.59	1.00
N = 356						

Key:
Cross-method, within-trait correlations are underlined.
Within-method, cross-trait correlations are **bold**.

evidence of what Campbell and Fiske call convergent validity, the agreement of different methods in measuring the same trait. Low correlations elsewhere (both within-method, cross-trait and cross-method, cross-trait) are indicative of discriminant validity. Of course, both convergent and discriminant validity are aspects of the assessment methods' construct validity. Neither of these two correlation matrices is displaying either high convergent validity or high discriminant validity. The method of assessment appears to be dominating the results, irrespective of what trait is being assessed.

To get a bit more enlightenment we will fit a measurement error model to the Cole *et al.* correlations. Quoting Cole *et al.* (1997):

> We allowed each measure to load onto exactly two factors, one trait factor and one method factor. For example, we allowed the depression self-report measure to load onto a depression factor and a self-report factor. We fixed all of the other loadings for this measure to be zero. Comparable loadings were allowed (and disallowed) for all of the other measures. Hence we extracted six factors in all: two trait factors (depression symptoms and anxiety symptoms) and four method factors (self-report, peer nomination, teacher rating, and parent report). We allowed the two trait factors to correlate with one another but not with the method factors. Likewise, we allowed the method factors to correlate with one another but not with the trait factors. The only additional assumption built into the model was that the loadings of measures onto the same method factor were constrained to be equal. (Trait factor loadings were not constrained).

Factors here correspond to the latent variables or characteristics we are trying to measure. The term 'loading' comes from the factor analysis literature and is equivalent to the regression coefficients (slope parameters) used in our measurement error models – see Section 1.4. Cole *et al.*'s model fits the data quite well – the authors reporting a chi-square statistic of 8.45 (with 9 degrees of freedom). Note that the model is being fitted to the covariance matrix (that is, we are ignoring intercepts in this analysis). Repeating their analysis using the summary statistics in Table 4.38(a) produced a chi-square of 7.94 – the difference presumably arising from the fact that the present analysis was based on published summary statistics given to only two decimal places. The present analysis yielded an estimate of the correlation between the two traits of 0.811, and estimates of the correlations between methods ranging from 0.158 to 0.339. We now introduce one further set of constraints: the correlations between method factors are *all* zero. The EQS program for fitting this model is given in Appendix 2(b). The resulting chi-square is 15.432 based on 15 degrees of freedom. The difference between the two chi-squares (15.432–7.94) is 7.492 with 6 degrees of freedom – clearly not statistically-significant. The estimates of the standardized loadings (slopes) are provided in Table 4.39. Note (in case of confusion) that, although the unstandardized slopes for the method factors are constrained to be equal across the two traits, this is not the case for the standardized values. The results are set out in Table 4.39 in order to clarify the model being fitted (we have abandoned the standard errors for simplicity). The notation (the labeling of the τs and δs, for example) should be clear from the data structure.

One further model tests whether depression and anxiety might be regarded as a unitary trait (i.e. a single one-dimensional characteristic). We replace the two trait factors by a single trait factor (equivalent to constraining the correlation between the two factors to be 1) and refit the model. The resulting chi-square is 16.852 with 16 degrees of freedom. Clearly the evidence that anxiety and depression can be measured separately in these children is very weak. The data in Table 4.38(a) refer to third-graders. Cole *et al.* also analysed data

Table 4.39 Parameter estimates from Depression/Anxiety correlations. (Data from Cole *et al.*, 1997).

Two common trait factors, four method factors
(Standardized results)

Depression
V1 – Self	$= 0.374\,\tau_D$	$+ 0.854\,\delta_S$	$+ 0.362\,\varepsilon_1$
V2 – Peers	$= 0.617\,\tau_D$	$+ 0.623\,\delta_{Pe}$	$+ 0.481\,\varepsilon_2$
V3 – Teacher	$= 0.501\,\tau_D$	$+ 0.713\,\delta_T$	$+ 0.491\,\varepsilon_3$
V4 – Parent	$= 0.530\,\tau_D$	$+ 0.780\,\delta_{Pa}$	$+ 0.333\,\varepsilon_4$

Anxiety
V5 – Self	$= 0.279\,\tau_A$	$+ 0.652\,\delta_A$	$+ 0.705\,\varepsilon_5$
V6 – Peers	$= 0.494\,\tau_A$	$+ 0.687\,\delta_{Pe}$	$+ 0.533\,\varepsilon_6$
V7 – Teacher	$= 0.430\,\tau_A$	$+ 0.778\,\delta_T$	$+ 0.458\,\varepsilon_7$
V8 – parent	$= 0.540\,\tau_A$	$+ 0.570\,\delta_{Pa}$	$+ 0.620\,\varepsilon_8$

$\mathrm{Corr}(\tau_D, \tau_A) = 0.962$.

from sixth-grade students and in this case the correlation between the two trait factors was lower (about 0.7) and significantly different from 1.

Returning to Table 4.38, we can infer that although each of the assessment methods is likely to be reasonably consistent (reliable) – as indicated by the sums of the squares of the standardized loadings for the appropriate trait and method loadings – the validities of the measures are all quite poor. The squares of the standardized trait loadings (slopes) are all pretty low compared to those of the method loadings and, of course, we have the problem of the near-perfect correlation between the trait factors for depression and anxiety.

Now let's consider a completely different area of application. Batista-Foguet, Coenders and Ferragud (2001) are concerned with the measurement of blood pressure in a survey of almost 6000 elderly patients. A physician used a mercury sphygmo-manometer to measure systolic blood pressure (SBP: trait 1) and diastolic blood pressure (DBP: trait 2). For each type of blood pressure measurement three readings were taken: in a seated position immediately on arrival at the doctor's office (method 1), in a seated position after a few minutes had elapsed (method 2) and in a standing position a few minutes later (method 3). The six measurements can be labeled SBP1, DBP1, SBP2, DBP2, SBP3 and DBP3. A general MTMM measurement model, as proposed by Batista-Foguet *et al.* (2001), for this data has the following structure:

$$
\begin{aligned}
\text{SBP1} &= \alpha_{11} + \beta_{11}\tau_1 + \gamma_{11}\delta_1 + \varepsilon_1 \\
\text{SBP2} &= \alpha_{12} + \beta_{12}\tau_1 + \gamma_{12}\delta_2 + \varepsilon_2 \\
\text{SBP3} &= \alpha_{13} + \beta_{13}\tau_1 + \gamma_{13}\delta_3 + \varepsilon_3 \\
\text{DBP1} &= \alpha_{21} + \beta_{21}\tau_2 + \gamma_{21}\delta_1 + \varepsilon_4 \\
\text{DBP2} &= \alpha_{22} + \beta_{22}\tau_2 + \gamma_{22}\delta_2 + \varepsilon_5 \\
\text{DBP3} &= \alpha_{23} + \beta_{23}\tau_2 + \gamma_{23}\delta_3 + \varepsilon_6
\end{aligned}
\tag{4.37}
$$

The two traits (τ_1 and τ_2) are allowed to be correlated but the method effects (δ_1, δ_2, and δ_3 are uncorrelated with each other and with the two method factors. The traits have means of μ_1 and μ_2, respectively, while the method effects all have zero mean. The slope parameters are the βs and γs, with αs being the intercepts. The εs are uncorrelated with each other and with all of the other the other latent variables. There are eleven different variances to be estimated from the data (for τ_1, τ_2, δ_1, δ_2, δ_3, ε_1, ε_2, ε_3, ε_4, ε_5 and ε_6). The model of Cole *et al.* for the childhood depression/anxiety data has a similar structure but with different constraints. Cole *et al.* comment that the assumption of equal method slopes was necessary for the structural identification of their model.

Identifiability constraints need to be introduced to (4.38). These could include, for example, $\alpha_{11} = \alpha_{21} = 0, \beta_{11} = \beta_{21} = 1$ and $\gamma_{11} = \gamma_{12} = \gamma_{13} = 1$. If we look at the expected covariance matrix after making these constraints, however, it is straightforward to show that there are still identifiability problems. These can be solved by the following constraints: $\gamma_{21} = \gamma_{22} = \gamma_{23}$. Problems of identification – including out-of-range estimates and convergence difficulties – are an important issue in the analysis of MTMM matrices (and many others!) and the interested reader is referred to Kenny and Kashy (1992) for a thorough discussion of these issues. Wothke (1996) provides a detailed review of MMTM methodology.

4.15 Longitudinal data

One of the purposes of this section is to introduce methods of analysis appropriate to the study of precision, reliability and stability in what social scientists call panel data (Markus, 1979). Panel data are typically thought of as information or measurements obtained from a fixed sample of correspondents (a panel) at two or more points in time. Each time point is referred to as a wave of the panel study. The number of measurements made at each time point is usually referred to as the number of indicators. Here we will concentrate on the use of multiple indicators at each of the waves of the panel study. These are discussed in detail in Wheaton *et al.* (1977), Jagodzinski *et al.* (1987) and by Raffalovich and Bohrnsdedt (1987). A detailed discussion of single indicator models can be found in Jagodzinski and Kühnel (1987).

First, consider a simple two-wave two-indicator panel study. At each of two distinct times two measurements are made on a panel of N subjects. Each subject thus provides four measurements from which a covariance or moments matrix can be calculated. The two measurements, for example, might be scores on two closely related reading tests administered when a panel of children are 10 years old (R_{11} and R_{12}) and again when they are aged 14 (R_{21} and R_{22}). Let us assume that the children's true reading abilities at 10 and 14 years are τ_1 and τ_2, respectively. A suitable measurement model might have the following form:

$$
\begin{aligned}
R_{11} &= \tau_1 + \varepsilon_1 \\
R_{12} &= \alpha_{12} + \beta_{12}\tau_1 + \varepsilon_2 \\
R_{21} &= \tau_2 + \varepsilon_3 \\
R_{22} &= \alpha_{22} + \beta_{22}\tau_1 + \varepsilon_4
\end{aligned}
\tag{4.38}
$$

with a non-zero correlation between τ_1 and τ_2. It is very likely that the two reading tests are performing similarly at the two ages. That is, $\alpha_{12} = \alpha_{22} = \alpha$, $\beta_{12} = \beta_{22} = \beta$, $\text{var}(\varepsilon_1) = \text{var}(\varepsilon_3) = \sigma_1^2$ and $\text{var}(\varepsilon_2) = \text{var}(\varepsilon_4) = \sigma_2^2$. This is an example of a very simple longitudinal factor analysis model (Raffalovich & Bohrnstedt, 1987).

In many measurement studies the model described by Equation (4.38) will not be adequate. Quite often a model is needed that incorporates equation errors that are subject-specific (here, child-specific) biases uncorrelated both with the true reading levels and with the measurement errors. As before, the subject-specific biases are assumed to have a zero mean. For a two-wave three-indicator (three reading tests, for example) data set the more realistic measurement model might have the following form:

$$R_{11} = \tau_1 + \delta_1 + \varepsilon_1$$
$$R_{12} = \alpha_2 + \beta_2\tau_1 + \delta_2 + \varepsilon_2$$
$$R_{13} = \alpha_3 + \beta_3\tau_1 + \delta_3 + \varepsilon_3$$
$$R_{21} = \tau_2 + \gamma_1\delta_1 + \varepsilon_4 \tag{4.39}$$
$$R_{22} = \alpha_2 + \beta_2\tau_2 + \gamma_2\delta_2 + \varepsilon_5$$
$$R_{23} = \alpha_3 + \beta_3\tau_2 + \gamma_2\delta_3 + \varepsilon_6$$

This is exactly analogous to the MTMM model described in Equation (4.37), and, like that model, it is not fully identified. The problem is again solved by forcing $\gamma_1 = \gamma_2 = \gamma_3$. More general longitudinal factor analysis models can be found in Raffalovich and Bohrnstedt (1987). Readers are also referred to a commentary on the Raffalovich and Bohrnstedt paper by Zeller (1987). In general, for these models to be identified without having to impose apparently arbitrary equality constraints, we need at least three indicators for each of at least three waves of measurement. One important characteristic of many panel studies is sample attrition. If subjects are lost from the study through a mechanism that is ignorable (MAR) then the methods discussed in Section 4.13 can be used to cope with this problem (the catch, however, is that we can rarely be sure about the drop-out mechanism).

A similar complication to that caused by attrition is the collection of data at a variable number of times; this variability being an inevitable characteristic of the trait(s) being measured rather than attrition as such. In Section 4.9 we mentioned a study by Oman *et al.* (1999) comparing two methods of measuring creatinine clearance (*MCC* and *ECC*) in cancer patients undergoing chemotherapy with a potentially nephrotoxic drug. They investigated the characteristics of the measurement methods in males and females separately. In the male file they had 184 pairs of measurements on 50 patients. The number of pairs of measurements was typically between 1 and 8, but there was one patient for which there were 12 pairs of measurements. A further complication of a data set such as this is that the time intervals between the repeated tests are not equal. In Section 4.9 we also mentioned that Bland and Altman (1999) describe a data set with a very similar structure, in which two methods of measuring cardiac output, impedence cardiography (IC) and radionuclide ventriculography (RV), were each used at regular intervals during surgery. The number of pairs of observations in this example was much less variable and ranged from 3 to 6. Letting $x = \log(MCC)$ and $y = \log(ECC)$, in order to attain normality, the equations for the jth pair of observations on the ith patient were specified by Oman *et al.* (1999) as follows:

$$x_{ij} = \tau + \varepsilon_{ij}$$
$$y_{ij} = \alpha + \beta\tau_i + \delta + \omega_{ij} \tag{4.40}$$

By now this model should be looking very familiar! The intercept and slope for method y are α and β, respectively. τ is the true value of the trait being measured, with mean μ_τ and variance σ_τ^2. The ε and ω are normally distributed random errors with zero mean and variance σ_ε^2 and σ_ω^2, respectively, and δ is a normally distributed patient-specific bias associated with method y. It has zero mean and variance σ_δ^2. Oman *et al.*, assumed that ε_{ij} and ω_{kl} were correlated when $i = k$ and $l = j$, but not otherwise. All other random variables were assumed to be independent. They then obtained point estimates for each of the model parameters using moments-based estimators, as described in Fuller (1987), and implemented bootstrap methods to obtain their standard errors. We refer the reader to the original paper for technical details.

4.16 Summary

In this chapter we have used a wide range of illustrative examples to demonstrate the main features of reasonably realistic measurement models. The take-home message is that a thorough investigation of the properties of competing methods of measurement, and of their measurement errors, is not always as simple as one might at first assume. In order to reveal these properties, however, we need reasonably informative designs. If we naïvely rely on pairs of unreplicated observations, as was the case in Chapter 3, then we will almost inevitably make naïve inferences and remain in the dark. The simple designs are inadequate for the job. If, however, we collect more informative data but fail to use the appropriate methodology to analyse these data, we will also stay in the dark. In order to progress we need informative (but not unnecessarily complex) designs together with a method of analysis that takes full advantage of the structure of the resulting data. We hope that the reader has been convinced that this is not too difficult an undertaking. Potentially, much can also be learnt through *Monte Carlo* simulation studies, at the design stage in particular, although we have not pursued this topic in any detail here. If nothing else, these simulation studies would inform the sample size calculations for the real thing.

5

Methods for categorical (binary) data

5.1 Introduction

So far, we have been concerned with the characteristics of quantitative measurements. In this, the final chapter, we turn our attention to categorical measurements and in particular, with binary (yes/no) indicators assumed to be assigning subjects or specimens to one of two underlying classes. These indicators are frequently referred to as binary diagnostic tests. As in the case of the quantitative measurements, we will be concerned primarily in situations in which we do not have access to an error-free 'gold standard' assessment and therefore the two underlying classes remain latent or hidden. The chapter begins with preliminary assessments of tests of relative bias and then the association or agreement between two or more binary assessments. The discussion then moves on to a description of analysis of variance methods for binary data and finally to the use of latent class models in the evaluation of binary diagnostic tests. Many of the concerns that have been raised in the case of the analysis of quantitative measurements will also be discussed with respect to latent class modelling. These include the identifiability of the models, the use of instrumental variables, replication, lack of conditional independence, multiple group analysis and analysis of results when there are data missing by design (as in two-phase designs). There is an enormous literature on the evaluation of diagnostic tests and there is no attempt in the present chapter to provide full details of all of this work, particularly that in which it is assumed that there is access to a 'gold standard' diagnosis. For this the following three texts might be consulted: Kraemer (1992), Zhou, Obuchowski & McClish (2002) and Sullivan Pepe (2003). Another area that barely gets a mention here is the assessment and modelling of agreement between multinomial (including ordinal) categorical measurements. This has been reviewed by Agresti (1992) and by Shoukri (2003).

5.2 Tests of relative bias

Table 5.1 shows the outcome of histological assessments of 118 biopsy slides by seven independent pathologists. The data are adapted from Holmquist *et al.* (1967) and have been reanalysed by Landis and Koch (1977b), Dunn (1989), Agresti (2002), and others. The

Table 5.1 Independent histological assessments by seven pathologists of the presence (1) or absence (0) of cancer of the cervix*.

Pathologist							Frequency
A	B	C	D	E	F	G	
0	0	0	0	0	0	0	34
0	0	0	0	1	0	0	2
0	1	0	0	0	0	0	6
0	1	0	0	0	0	1	1
0	1	0	0	1	0	0	4
0	1	0	0	1	0	1	5
1	0	0	0	0	0	0	2
1	0	1	0	1	0	1	1
1	1	0	0	0	0	0	2
1	1	0	0	0	0	1	1
1	1	0	0	1	0	0	2
1	1	0	0	1	0	1	7
1	1	0	0	1	1	1	1
1	1	0	1	0	0	1	1
1	1	0	1	1	0	1	2
1	1	0	1	1	1	1	3
1	1	1	0	1	0	1	13
1	1	1	0	1	1	1	5
1	1	1	1	1	0	1	10
1	1	1	1	1	1	1	16
					Total		118

* Adapted from Holmquist *et al.* (1967).

study was designed to investigate the variability in the histological classification of carcinoma *in situ*, and other related lesions of the uterine cervix. The original coding used by the seven pathologists was: (1) negative, (2) atypical squamous hyperplasia, (3) carcinoma *in situ*, (4) squamous carcinoma with early stromal invasion, and (5) invasive carcinoma. Following Landis and Koch (1997b) the categories have been combined in the following way: (0) derived from the original categories (1) and (2), and (1) derived from the original categories (3)–(5). The two new categories correspond to the presence or absence of carcinoma. The proportions of the 118 slides perceived to be displaying carcinoma by each of the seven pathologists (A to G in alphabetical order) are:

$$0.559, 0.669, 0.381, 0.271, 0.602, 0.212 \text{ and } 0.559$$

Pathologists A, B, E and G appear to have reasonably similar marginal proportions (average 0.597). Pathologists C, D and F also show some similarities (average 0.288). Here our interest is in testing whether these differences between the marginal proportions for the seven pathologists are significantly different and if so, where the differences might lie. The appropriate test of marginal homogeneity across the seven observers is based on the use of Cochran's Q statistic (Fleiss, 1965).

Let X_{ik} represent the assessment of the ith slide by the kth pathologist ($i = 1$ to I, $k = 1$ to K) with $X_{ik} = 1$ if the ith slide is judged by the kth observer to be indicative of carcinoma, 0 otherwise. Let X_{i+} represent the total number of observers who judge the ith slide to be indicative of carcinoma and, similarly, let X_{+k} represent the total number of slides the kth

Table 5.2 Two-way contingency table for assessment of bias and agreement between two binary measurements.

Observed counts (proportions)

		Rater 2 category		
		0	1	Total
Rater 1	0	n_{00} (p_{00})	n_{01} (p_{01})	n_{0+} $p_{0+} = n_{0+}/n$
category	1	n_{10} (p_{10})	n_{11} (p_{11})	n_{1+} $p_{1+} = n_{1+}/n$
	Total	n_{+0}	n_{+1}	n
		$p_{+0} = n_{+0}/n$	$p_{+1} = n_{+1}/n$	

pathologist judges to have indication of carcinoma. Finally, let X_{++} be the total number of slides for which carcinoma is judged to be present. Cochran's Q statistic is given by

$$Q = \frac{K(K-1)\sum_{k=1}^{k}\left(X_{+k} - X_{++}/K\right)^2}{KX_{++} - \sum_{i=1}^{I} X_{i+}^2} \tag{5.1}$$

If the null hypothesis of marginal homogeneity is true, then for large samples, Q will be approximately distributed as a chi-square with $K-1$ degrees of freedom. For the data given in Table 5.1, $Q = 181.59$ with 6 degrees of freedom. In the case of just two raters (i.e. $K = 2$) Cochran's Q test is equivalent to the well-known McNemar test for the comparison of two correlated proportions (McNemar, 1947).

More general methods of comparing the marginal distributions of categorical assessments made by two or more raters are described by Landis and Kock (1977a). These are based on theoretical work explained in detail in Grizzle et al. (1969) or Appendix 1 of Koch et al. (1977). An alternative, which is particularly straightforward for binary assessments, is the use of marginal logistic regression models and generalized estimating equations (GEE) methodology (Liang & Zeger, 1986; Zeger & Liang, 1986).

Having established that there are highly significant differences in the proportions of pathologists identifying carcinoma, we can move on to look at differences between each pair of pathologists (or, instead, we could think of a priori contrasts between groups of pathologists but there is no justification for doing so in this case). Table 5.2 sets out a standard notation for a general 2×2 contingency table for the comparison of binary assessments made by each of two raters. Using this notation, McNemar's test statistic for the comparison of the marginal probabilities (p_1 and p_2) is given by

$$X^2 = \frac{\left(n_{01} - n_{10}\right)^2}{n_{01} + n_{10}} \tag{5.2}$$

Under the null hypothesis of marginal homogeneity this will have an asymptotic distribution as a chi-square with one degree of freedom. Its calculation is illustrated in the case of pathologists A and B in Table 5.3. Here the difference is clearly statistically significant (using a significance level, α, of 0.05) but if we are going to examine all of the possible 21 pairwise comparisons of the seven pathologists in Table 5.1, perhaps we would be safe to use some sort of adjustment to our chosen significance level. If we use the familiar Bonferroni adjustment

Table 5.3 Joint assessment of biopsy slides by pathologists A and B (see Table 5.1).

		Pathologist B		
		0	1	Total
Pathologist A	0	36	16	52
	1	3	63	66
	Total	39	79	118

McNemar's test: $X^2 = (16 - 3)^2/(16 + 3) = 8.895$

Table 5.4 McNemar X^2 statistics for the 21 paired comparisons of the seven pathologists in Table 5.1*.

	Pathologist						
	B	E	A	G	C	D	F
B	–						
E	4.57	–					
A	8.89	1.47	–				
G	11.27	2.27	0	–			
C	32.11	26.00	21.00	21.00	–		
D	47.00	37.10	34.00	34.00	6.76	–	
F	54.00	46.00	41.00	41.00	14.29	2.56	–

* The pathologists have been re-ordered according to the overall proportion of biopsy slides perceived to be showing evidence of carcinoma (from B, the highest, to F, the lowest).

(see Fleiss, 1986) then we divide our nominal significance level by the number of comparisons being made (so for the present example the revised significance level, α^*, is $0.05/21 = 0.0024$ – with a corresponding critical value for a single degree of freedom chi-square of about 9.5). The results of these 21 comparsions are summarized in Table 5.4. Clearly many of them are indeed statistically significant when using an adjusted significance level of 0.0024. Note that these statistics again indicate that there might be two distinct groups of pathologists (A, B, E and G versus C, D and F).

5.3 Association, concordance and reliability

In the analysis of data from method comparison studies using quantitative measurement procedures, the obvious place to start is to have a look at scatter diagrams and patterns of correlation between the pairs of methods. Here we start with correlations (called ϕ coeffcient in the case of binary data). These are shown in Table 5.5(a). For comparison, Table 5.5(b) shows the corresponding values for Cohen κ. Cohen (1960) suggested this statistic as a measure of chance-corrected agreement and, using the notation of Table 5.2, it is defined as follows:

$$\kappa = \frac{(p_{00} + p_{11}) - (p_{0+}p_{+0} + p_{1+}p_{+1})}{1 - (p_{0+}p_{+0} + p_{1+}p_{+1})}$$

$$= \frac{P_{Observed} - P_{Chance}}{1 - P_{Chance}} \tag{5.3}$$

where $P_{Observed}$ is the proportion of observed agreements between the two raters and P_{Chance} is that expected by chance after allowing for the raters' different margins. The pattern of κ

Table 5.5 Associations and concordance between pathologists' biopsy ratings*.

(a) Correlation (ϕ coefficient)

	Pathologist						
	B	E	A	G	C	D	F
B	–						
E	0.75	–					
A	0.68	0.71	–				
G	0.76	0.81	0.79	–			
C	0.51	0.64	0.70	0.70	–		
D	0.43	0.46	0.54	0.54	0.54	–	
F	0.36	0.42	0.46	0.46	0.49	0.57	–

(b) Cohen's κ coefficient

	Pathologist						
	B	E	A	G	C	D	F
B	–						
E	0.75	–					
A	0.66	0.70	–				
G	0.74	0.81	0.79	–			
C	0.44	0.58	0.65	0.65	–		
D	0.31	0.36	0.45	0.45	0.52	–	
F	0.23	0.30	0.35	0.35	0.45	0.56	–

(c) Intra-class correlation (intra-class κ)

	Pathologist						
	B	E	A	G	C	D	F
B	–						
E	0.75	–					
A	0.66	0.71	–				
G	0.73	0.81	0.80	–			
C	0.39	0.56	0.65	0.65	–		
D	0.20	0.30	0.41	0.41	0.52	–	
F	0.08	0.20	0.27	0.27	0.43	0.56	–

*The pathologists have been re-ordered according to the overall proportion of biopsy slides perceived to be showing evidence of carcinoma (from B, the highest, to F, the lowest).

values is very similar to those for ϕ. Although κ is usually regarded as a measure of agreement (concordance), it is really just an indicator of association. A better option for a measure of concordance is the intra-class correlation coefficient (identical to Kraemer's intra-class κ – see Kraemer, 1979; Kraemer et al., 2002). The latter can be calculated and interpreted as a chance-corrected measure of agreement (as in Equation 5.3 but after assuming marginal homo-geneity for the raters, and therefore replacing the individual rater's marginal proportions by a common estimate) and is equivalent to Scott's π coefficient (Scott, 1955). As in the case for quantitative measurements, there are several ways of actually calculating the intra-class correlation (Ridout et al., 1999) and those in Table 5.5(c) have been derived via a one-way analysis of variance (see the following section). Another possible candidate for a measure of concordance between paired measurements is the categorical version of the Krippendorff–Lin coefficient (Section 2.3; see also Krippendorff, 1970 and Schouten 1985, 1986). It would, however, give very similar results to those for the intra-class correlation.

Table 5.6 One-way analysis of variance for pathologist's ratings (restricting the analysis to pathologists A, B, E and G).

Source of variation	Sum of squares	d.f.	Mean square
Between subjects	91.517	117	0.7822
Within subjects	22.000	354	0.0621
Intra-class correlation = (0.782 − 0.062)/(0.782 + 3 × 0.062) = 0.74			

Note that the agreement indices towards the bottom left corner of Table 5.5(c) are very low, but those for Pathologists B, E, A and G appear to form an internally consistent group of observers.

Let's, for the time being, restrict our analysis to Pathologists A, B, E, and G, and assuming that they have the same characteristics as measuring 'instruments', consider how we might estimate their common reliability (s.e. defined by Kraemer, 1979 – see Section 1.7). One possibility is to simply average the six intra-class τs given in the top rows of Table 5.5(c) (see Light, 1981; Conger, 1980 and Kraemer *et al.*, 2002). The average value of these six κs (0.75, 0.66, 0.71, 0.73, 0.81, 0.80) is 0.74. This is an estimate of the reliability of an assessment made by a randomly selected pathologist (from the four under consideration) of a randomly selected biopsy sample. If we carry out the usual one-way analysis of variance (that is, assessment by subject) using the data from these four pathologists together, we obtain the results given in Table 5.6. The estimate for the intra-class correlation derived from the mean squares from this ANOVA table is again 0.74. Note that we report neither the F-statistics nor the P-values which would usually be associated with the ANOVA table. They are based on normality assumptions which clearly do not hold for binary data. Having introduced ANOVA models for binary data, we will discuss them in further detail in the following section. We will then return to consider inferential procedures for intra-class correlations and variations of the κ coefficient in Section 5.5.

5.4 ANOVA models for categorical measurements

Returning to Table 5.1, let's select two disparate pathologists (A and B, for example) and analyse the 118 pairs of assessments provided by these observers in further detail. Table 5.3 provides the basic table of cell counts; Table 5.7 provides the results of (a) a one-way ANOVA and (b) a corresponding two-way ANOVA.

From the one-way ANOVA results we obtain an intra-class correlation estimate of 0.663. For a reasonably large number of subjects (biopsies), the standard moments based estimator from the ANOVA table (i.e. Equation 2.29) is the same as

$$R_1 = \frac{S - (O + E)}{S + (O + E)} \tag{5.4}$$

in which O is the sum of squares due to raters (pathologists), S is the sum of squares due to subjects (biopsies) and finally, E is the error sum of squares. Substituting the values given in Table 5.7(a) for these sums of squares we obtain

$$R_1 = \frac{46.411 - (0.7161 + 8.7839)}{46.411 + (0.7161 + 8.7839)} = 0.660 \tag{5.5}$$

Table 5.7 Analysis of variance (ANOVA) for pathologists A and B.

(a) One-way ANOVA

Source of variation	Sum of squares	d.f.	Mean square
Between subjects	46.411	117	0.3967
Within subjects	9.500	118	0.0805

Intra-class correlation $= (0.3967 - 0.0805)/(0.3967 + 0.0805) = 0.663$

(b) Two-way ANOVA

Source of variation	Sum of squares	d.f.	Mean square
Between subjects	46.411	117	0.3967
Between pathologists	0.7161	1	0.7161
Residual (error)	8.7839	117	0.0751

Using the notation for the cell and marginal counts given in Table 5.2, it is straightforward to show that R_1 is also given by

$$R_1 = \frac{4(n_{00}n_{11} - n_{01}n_{10}) - (n_{01} - n_{10})^2}{(n_{0+} + n_{+0})(n_{1+} + n_{+1})} = 0.660 \tag{5.6}$$

Equation (5.6) is an alternative expression to Scott's π, the chance-corrected coefficient of agreement introduced in Section 5.3. The equivalence of Scott's π and R_1 was demonstrated by Fleiss (1975). The assumption of marginal homogeneity in the calculation of Scott's π coefficient is equivalent to that of ignoring rater (pathologist) biases and using a one-way ANOVA model for the derivation of an intra-class correlation coefficient.

If we were not prepared to assume the lack of rater bias, then the appropriate model would be either be a two-way random effects ANOVA or a two-way mixed model. It will be assumed there is no biopsy specimen-pathologist interaction effect. Taking the random effects model first, the appropriate intra-class correlation can be estimated by

$$\hat{\rho}_{2WR} = \frac{\hat{\sigma}_S^2}{\hat{\sigma}_S^2 + \hat{\sigma}_P^2 + \hat{\sigma}_E^2} \tag{5.7}$$

where $\hat{\sigma}_S^2$, $\hat{\sigma}_P^2$, and $\hat{\sigma}_E^2$, are estimated variance components for specimens (subjects), pathologists (raters) and error (residual), respectively. Here

$$\sigma_{subj}^2 = \frac{(0.3967 - 0.0751)}{2} = 0.1608$$

$$\hat{\sigma}_{path}^2 = \frac{(0.7161 - 0.0751)}{118} = 0.00543$$

$$\hat{\sigma}_{error}^2 = 0.0751$$

and, therefore,

$$\hat{\rho}_{2WR} = \frac{0.1608}{(0.1608 + 0.00543 + 0.0751)} = 0.666$$

Fleiss (1975) demonstrates that (5.7) is approximately the same as

$$R_2 = \frac{S - E}{S + E + 2O} \tag{5.8}$$

where S, O and E are the same sums of squares defined for Equation (5.4). For the present example

$$R_2 = \frac{46.411 - 8.7839}{46.411 + 8.7839 - 2 \times 0.7161} = 0.664$$

Alternative expression to (5.8) is given by

$$R_2 = \frac{2(n_{00}n_{11} - n_{01}n_{10})}{(n_{0+}n_{+1} + n_{+0}n_{1+})} = 0.664 \tag{5.9}$$

Equation (5.9) is an equivalent to Cohen's κ as defined by Equation (5.3).

Finally, considering a mixed effects ANOVA, the estimated intra-class correlation for a single rater (pathologist) is given by

$$\hat{\rho}_{2WR} = \frac{\hat{\sigma}_S^2}{\hat{\sigma}_S^2 + \hat{\sigma}_E^2} = 0.682 \tag{5.10}$$

which, in turn, can be shown to be identical to

$$R_3 = \frac{2(n_{00}n_{11} - n_{01}n_{10})}{(n_{0+}n_{1+} + n_{+0}n_{+1})} \tag{5.11}$$

The coefficient defined by (5.11) is equivalent to the statistic r_{11} described by Maxwell and Pilliner (1968).

5.5 ANOVA models for more than two raters

We have already briefly introduced the one-way ANOVA model for multiple raters in Section 5.3. The extension and generalization of the Cohen κ coefficient (equivalent to the intra-class correlation from a two-way random effects ANOVA model) for more than two raters has been discussed by several authors (see, for example, Fleiss, 1971; Light, 1971; Hubert, 1977b, Conger, 1980; Davies and Fleiss, 1982; Rae, 1984 and 1988; Schouten, 1986; and Janson & Olsson, 2001). Here we essentially follow the approach of Davies and Fleiss (1982). Despite having said in the introduction to this chapter that we will not discuss multinomial categorical data we will, in fact, consider the case in which each of I subjects is classified into C ($C \geqslant 2$) mutually exclusive categories by each of the same set of K observers or raters. We will then deal with binary ratings ($C = 2$) as a special case.

Let the observation for the ith subject ($i = 1, 2, ..., I$) by the kth rater ($k = 1, 2, ..., K$) be represented by the vector

$$X_{ik} = (X_{ik1}, X_{ik2}, ..., X_{ikC})'$$

where each X_{ikc} ($c = 1, 2, ..., C$) has the value 0 or 1. If the rating or grade is actually c then $X_{ikc} = 1$, otherwise $X_{ikc} = 0$. The vector X_{ik}, therefore, will have only one element with the value 1 and C–1 elements with the value 0 (implying that $\sum_{c=1}^{C} X_{ikc} = 1$ for all values of both i and k.

Now, for the ith subject let

$$Y_{ic} = \sum_{K=1}^{C} X_{ikc}$$

That is, Y_{ic} is the number of raters or observers who allocate the ith subject to the rating or category c. It follows that

$$\sum_{c=1}^{C} Y_{ic} = \sum_{k=1}^{K} \sum_{c=1}^{C} X_{ikc} = K$$

For each subject there are a total of $\frac{1}{2}K(K-1)$ pairs of classifications. The observed number of pairs of observations for this subject that are in agreement is

$$\frac{1}{2} \sum_{c=1}^{C} Y_{ic}(Y_{ic} - 1)$$

Adding over the I subjects, the observed proportions of possible pairs of classifications that are in agreement is given by

$$
\begin{aligned}
P_{Obs} &= \frac{1}{IK(K-1)} \sum_{i=1}^{I} \sum_{c=1}^{C} Y_{ic}(Y_{ic} - 1) \\
&= \frac{1}{IK(K-1)} \sum_{i=1}^{I} \sum_{c=1}^{C} Y_{ic}(Y_{ic}^2 - IK)
\end{aligned}
\tag{5.12}
$$

Let π_{kc} be the probability that a randomly chosen subject is placed into category c by rater k. This probability is estimated by

$$p_{ik} = \frac{1}{I} \sum_{i=1}^{I} X_{ikc} \tag{5.13}$$

This is simply the observed proportion of the I subjects placed into category c by the kth rater. If the classifications by raters are statistically independent, then the probability that a given pair or raters (k and m, say) will agree in the classification of a randomly selected subject is estimated by

$$\sum_{c=1}^{C} p_{kc} p_{mc}$$

The average chance-expected probability of pairwise agreement is then estimated by

$$P_{Chance} = \frac{1}{K(K-1)} \sum_{k=1}^{K} \sum_{m=1, \, m \neq k}^{K} \sum_{c=1}^{C} p_{kc} p_{mc} \tag{5.14}$$

Finally, as in (5.3), κ is defined by

$$\kappa = \frac{P_{Obs} - P_{Chance}}{1 - P_{Chance}} \tag{5.15}$$

Equation (5.15) can be shown (Davies & Fleiss, 1982) to be equivalent to

$$\kappa = 1 - \frac{IK^2 - \sum_{i=1}^{I}\sum_{c=1}^{C} Y_{ic}^2}{I\left\{K(K-1)\sum_{c=1}^{C} \bar{p}_c(1-\bar{p}_c) + \sum_{c=1}^{C}\sum_{k=1}^{K}(p_{kc} - \bar{p}_c)^2\right\}} \tag{5.16}$$

where $\bar{p}_c$ is the average (over raters) of the p_{kc}. That is,

$$\bar{p}_c = \frac{1}{K}\sum_{k=1}^{K} p_{kc}$$

There are two interesting special cases of Equation (5.16). The first occurs when it can be assumed that the p_{kc} are the same for all values of k (that is, marginal homogeneity). In this case (5.16) simplifies to

$$\kappa = 1 - \frac{IK^2 - \sum_{i=1}^{I}\sum_{c=1}^{C} Y_{ic}^2}{I\left\{K(K-1)\sum_{c=1}^{C} \bar{p}_c(1-\bar{p}_c)\right\}} \tag{5.17}$$

The second interesting special case is when there are only two possible classifications (that is, $C = 2$). Here equation (5.16) simplifies to

$$\kappa = 1 - \frac{\sum_{i=1}^{I} Y_{i2}(K - Y_{i2})}{I\left\{K(K-1)\bar{p}_2(1-\bar{p}_2) + \sum_{k=1}^{K}(p_{k2} - \bar{p}_2)^2\right\}} \tag{5.18}$$

Equation (5.18) is, itself, equivalent to

$$\kappa = \frac{MSS - MSE}{MSS + (K-1)MSE + K(MSO)/(I-1)} \tag{5.19}$$

where MSS, MSO and MSE are the mean squares due to subjects, observers and measurement error, respectively. These mean squares are derived from a two-way analysis of variance of the binary ratings (Davies & Fleiss, 1982). The corresponding moments estimator for the intra-class correlations derived from a two-way analysis of variance is given by Bartko (1966).

$$R = \frac{MSS - MSE}{MSS + (K-1)MSE + K(MSO - MSE)/I} \tag{5.20}$$

In practice, the distinction between (5.19) and (5.20) is of little consequence. Using the ANOVA table provided in Table 5.8, for the pathologists' data (using all seven pathologists) we obtain 0.520 and 0.522, respectively.

Table 5.8 Two-way analysis of variance (ANOVA) for all seven pathologists.

Source of variation	Sum of squares	d.f.	Mean square
Between subjects	119.482	117	1.021
Between pathologists	22.058	6	3.676
Residual (error)	63.942	702	0.091

Table 5.9 Pairwise κ coefficients for the seven pathologists in Table 5.1.

Pair	P_{Obs}	P_{Chance}	$P_{Obs} - P_{Chance}$	$1 - P_{Chance}$	κ
AB	0.8390	0.5201	0.3189	0.4799	0.6645
AC	0.8220	0.4859	0.3361	0.5141	0.6538
AD	0.7119	0.4729	0.2390	0.5271	0.4534
AE	0.8559	0.5121	0.3438	0.4879	0.7047
AF	0.6525	0.4658	0.1867	0.5342	0.3497
AG	0.8993	0.5070	0.3913	0.4930	0.7937
BC	0.6949	0.4598	0.2351	0.5402	0.4352
BD	0.6017	0.4224	0.1793	0.5776	0.3104
BE	0.8814	0.5345	0.3469	0.4655	0.7452
BF	0.5424	0.4023	0.1401	0.5977	0.2346
BG	0.8729	0.5201	0.3528	0.4799	0.7352
CD	0.7881	0.5543	0.2338	0.4457	0.5246
CE	0.7797	0.4759	0.3038	0.5241	0.5797
CF	0.7627	0.5684	0.1943	0.4316	0.4502
CG	0.8220	0.4859	0.3361	0.5141	0.6538
DE	0.6525	0.4535	0.1990	0.4565	0.3641
DF	0.8390	0.6319	0.2071	0.3681	0.5626
DG	0.7119	0.4729	0.2390	0.5271	0.4534
EF	0.6102	0.4414	0.1688	0.5586	0.3022
EG	0.9068	0.5121	0.3947	0.4879	0.8090
FG	0.6525	0.4658	0.1867	0.5342	0.3495
Total			5.5333	10.6350	

We finish by pointing out the relationship between the κ coefficients defined by (5.16) and the pairwise κs displayed in Table 5.5(b). For each pair of raters, k and m, $(k \neq m)$ define a κ coefficient as

$$\kappa^{km} = \frac{P^{km}_{Obs} - P^{km}_{Chance}}{1 - P^{km}_{Chance}} \tag{5.21}$$

where P^{km}_{Obs} and P^{km}_{Chance} are the observed and chance expected proportions of agreement, respectively. These values are given in Table 5.9 for the seven pathologists in Table 5.1. Note from Table 5.9, that, summing over all pairs of raters

$$\frac{\sum \left(P^{km}_{Obs} - P^{km}_{Chance} \right)}{\sum \left(1 - P^{km}_{Chance} \right)} = 0.520$$

This is the κ coefficient calculated using Equation (5.19). It is a weighted average of the individual κ^{km}s as follows:

$$\kappa = \frac{\sum \left(1 - P^{km}_{Chance} \right) \kappa^{km}}{\sum \left(1 - P^{km}_{Chance} \right)} \tag{5.22}$$

Again, the summation is carried out over all 21 different pathologist pairs (see Schouten, 1985). In comparison, the arithmetic mean of the matrix of pairwise kappas given in Table 5.5(b) is 0.530. There is little difference between the two.

In none of our discussions of the various intra-class correlations or κ coefficients have we discussed any form of statistical inference (significance testing, standard error estimation or confidence interval estimation, for example). This is because we are primarily concerned with the use of these coefficients as summary statistics (primarily measures of association rather than agreement) which may be useful in an initial exploration of the data (Bloch & Kraemer, 1989; Kraemer et al., 2002). Although many authors use them as a basis of various models or agreement or reliability and consider the estimation of the parameters of such models in a more formal manner, this is not the direction that will be taken here. Maximum likelihood estimation of κ coefficients can be found, for example, in Bishop, Fienberg & Holland (1975), Bloch & Kraemer (1989), Bodian (1994), and in Klar et al. (2002). The latter authors consider interval estimation for κ, as do, for example, Hale & Fleiss (1993), Lui & Kelly (1999), Blackman & Koval (2000), Nam (2000), Altaye, Donner & Klar (2001) and Altaye, Donner & Eliasziw (2001). Davies and Fleiss (1982) provide an approximate standard error to the ANOVA-based estimator in Equation (5.16) and several authors have suggested the use of the jack-knife or bootstrapping for the estimation of the standard errors of various κs (see, for example, Schouten, 1986; Dunn, 1989; Kraemer et al., 2002; and Klar et al., 2002). See also Lipsitz et al. (1994), Koval & Blackman (1996), Klar et al. (2000) and Thompson (2001). Investigators contemplating the use of the jack-knife or boot-strapping should bear in mind the sampling design used to generate the data (see Section 2.7, for further discussion).

5.6 Latent class models

Let's return to the data for Pathologists A and B in Table 5.3. What might be an underlying statistical or measurement model for these data? One plausible approach is that, although it is not known, there is a true diagnostic status for each biopsy slide (either the tissue from which it was obtained contained cells typical of carcinoma, or it did not). So, we start by defining this latent class indicator by $D = 1$ if the disease (carcinoma) is present, and $D = 0$, otherwise. We also define the underlying proportion of individuals with carcinoma (prevalence) in the population from which the biopsies were taken to be π. Now, given that carcinoma is present ($D = 1$) the sensitivity of Pathologist A is the proportion of these diseased slides that in the long run would be detected by this Pathologist. Let $T_A = 1$ if Pathologist A thinks that carcinoma is present, $T = 0$ otherwise. Then the sensitivity of Pathologist A is given by

$$\alpha_A = \text{Prob}(T_A = 1 \mid D = 1) \tag{5.23}$$

The specificity of Pathologist A is the proportion of the slides without carcinoma that the pathologist correctly identifies as being disease-free. That is,

$$\beta_A = \text{Prob}(T_A = 0 \mid D = 0) \tag{5.24}$$

Similarly, for Pathologist B,

$$\alpha_B = \text{Prob}(T_B = 1 \mid D = 1) \tag{5.25}$$

and

$$\beta_B = \text{Prob}(T_B = 0 \mid D = 0) \tag{5.26}$$

Table 5.10 A simple latent class model: expected results of the ratings of biopsies by Pathologists A and B*.

Outcome		Joint probability
Pathologist A	Pathologist B	
0	0	$\pi(1 - \alpha_A)(1 - \alpha_B) + (1 - \pi)\beta_A\beta_B$
0	1	$\pi(1 - \alpha_A)\alpha_B + (1 - \pi)\beta_A(1 - \beta_B)$
1	0	$\pi\alpha_A(1 - \alpha_B) + (1 - \pi)(1 - \beta_A)\beta_B$
1	1	$\pi\alpha_A\alpha_B + (1 - \pi)(1 - \beta_A)(1 - \beta_B)$

* See text for definitions of the parameters.

Now, if we can safely assume that given the true disease status (D) the ratings of the two pathologists are statistically independent, then the probability of seeing the data in each of the cells of Table 5.3 is given in Table 5.10. The joint likelihood for the counts in Table 5.3 is given by

$$L = \left[\pi(1 - \alpha_A)(1 - \alpha_B) + (1 - \pi)\beta_A\beta_B\right]^{n_{11}}\left[\pi(1 - \alpha_A)\alpha_B + (1 - \pi)\beta_A(1 - \beta_B)\right]^{n_{01}}$$
$$\times \left[\pi\alpha_A(1 - \alpha_B) + (1 - \pi)(1 - \beta_A)\beta_B\right]^{n_{10}}\left[\pi\alpha_A\alpha_B + (1 - \pi)(1 - \beta_A)(1 - \beta_B)\right]^{n_{00}}$$

(5.27)

In principal this likelihood could be maximized to obtain estimates of the model parameters. But we have a problem. The model is not identified. We have five unknowns (π, α_A, β_A, α_B and β_B) but only three degrees of freedom. We are in an analogous situation to that in a method comparison study involving two unreplicated quantitative methods (Chapter 3) and the solutions are essentially similar. We can constrain some of the parameters; we can find an instrumental variable; we can replicate one or both of the ratings; or we can find one or more new methods (none of these solutions are mutually exclusive, of course). Without further data the only option is the introduction of parameter constraints. Three possibilities are the following (Zhou *et al.*, 2002):

(a) Assume that the sensitivity and specificity for one of the pathologists is known (from previous work, for example).
(b) Assume that the sensitivities (the αs) of both pathologists are equal to 1.
(c) Assume that the specificities (the βs) of both pathologists are equal to 1.

An alternative to the introduction of constraints is the use of prior information together with Bayesian methods (see Zhou *et al.*, 2002; Johnson & Gastwith, 1991; and Joseph *et al.*, 1995; Dendukuri & Joseph, 2001), but, like Andersen (1997), I do not find this approach entirely convincing. Here we choose to find more data.

5.7 Latent class models with three or more test methods

If, as in Table 5.1, we have more than two conditionally independent test methods (here pathologists) then the resulting model is identified. It is just identified when the number of methods is three (that is, there are no degrees of freedom with which to check the fit of the

Table 5.11 Latent class model parameter estimates (with standard errors) for the Holmquist biopsy data.

Prevalence	Estimate	Sensitivity	Estimate (s.e.)	Specificity	Estimate (s.e.)
π	0.501	α_A	1.000 (0.000)	β_A	0.883 (0.043)
		α_B	0.983 (0.017)	β_B	0.646 (0.064)
		α_C	0.761 (0.057)	β_C	1.000 (0.000)
		α_D	0.541 (0.065)	β_D	1.000 (0.000)
		α_E	0.979 (0.021)	β_E	0.777 (0.055)
		α_F	0.423 (0.065)	β_F	1.000 (0.000)
		α_G	1.000 (0.000)	β_G	0.883 (0.117)

model – again an analogous situation to the comparison of three quantitative methods as in Section 4.4). The model is over-identified with four or more test methods (that is, there are degrees of freedom left over to test the fit of the model) and we can use this to ask whether we need to relax the conditional independence assumptions described in the previous Section.

Typically (but not necessarily) maximum likelihood estimates will be obtained through the use of the EM algorithm and standard errors of the parameter estimates then obtained from the inverse of approximations to the Fisher information matrix (see Bartholomew & Knott, 1999; Muthén & Muthén, 1998–2001). The failure to obtain such an inverse is often an indicator of lack of identifiability of the model. It is possible to test equality constraints across tests (methods) and also simultaneously to fit latent class models to two or more samples of data (for the latter, see Clogg & Goodman, 1985). Fitting a latent class model to the biopsy data (using *MPlus* – see Appendix 5(a)) gives the estimates listed in Table 5.11. Note that some of the estimates are equal to 1 (on the boundary of the permitted parameter interval) and have a reported standard error of 0. On the basis of these estimates, the pathologists seem to fall into the same two distinct groups as before (A, B, E and G versus C, D and F). The expected counts obtained through the use of the parameter estimates are given in Table 5.12.

At the bottom of Table 5.12 is given the values of two goodness-of-fit statistics, that is, the Pearson chi-square,

$$X^2 = \sum_{\text{cells}} \frac{(\text{Observed} - \text{Expected})^2}{\text{Expected}} \tag{5.28}$$

and the likelihood ratio chi-square,

$$G^2 = 2 \sum_{\text{cells}} \text{Observed} \times \log\left(\frac{\text{Observed}}{\text{Expected}}\right) \tag{5.29}$$

Asymptotically, both of these statistics are distributed as χ^2 with degrees of freedom

$$\nu = 2^K - q - 1 \tag{5.30}$$

where K is the number of tests (pathologists) being administered and q is the number of estimated parameters. For the pathologists' biopsy data, $K = 7$ and $q = 15$ (the estimated parameters including 7 sensitivities, 7 specificities and the disease prevalence), hence $\nu = 128 - 15 - 1 = 112$. Neither X^2 nor G^2 is indicating any significant lack of fit (in fact the fit as indicated by these two statistics looks too good, particularly if we compare observed and expected counts in the table) but they are very different from each other.

Table 5.12 Observed and expected counts after fitting latent class model for the Holmquist biopsy data.

Pattern	Observed count	Expected count	Chi-square contribution	
			Pearson	Likelihood
0000000	34	23.05	5.20	26.43
1000000	2	3.04	0.36	−1.67
0100000	6	12.65	3.50	−8.95
1100000	2	1.67	0.07	0.73
0000100	2	6.61	3.22	−4.78
0100100	4	3.63	0.04	0.78
1100100	2	0.48	4.84	5.72
1000001	1	1.67	0.27	−1.02
1100001	1	0.30	1.65	2.42
1101001	1	0.09	8.87	4.76
0100101	5	0.48	42.72	23.46
1100101	7	3.67	3.03	9.05
1010101	1	0.20	3.27	3.25
1110101	13	11.47	0.20	3.25
1101101	2	4.25	1.19	−3.02
1111101	10	13.35	0.92	−6.04
1100111	1	2.64	1.02	−1.94
1110111	5	8.40	1.38	−5.19
1101111	3	3.11	0.00	−0.22
1111111	16	9.90	3.76	15.36
Sub-total			85.48	62.37
Contribution from empty cells			7.17	0.00
Total			92.65	62.37

Why is this? The problem arises from the fact that we are fitting a model to a very sparse contingency table (of the $2^7 = 128$ possible response patterns, 108 have an observed count of zero, with a further 11 having a count lower than five). It is clear from Equation (5.29) that response patterns with an observed count of zero do not contribute to G^2, regardless of the count predicted by the model. This partly explains why G^2 is likely to be lower than X^2 for a sparse contingency table. But the problem is that the asymptotic properties of both of these two statistics breaks down for data such as those in Table 5.12. A *Monte Carlo* simulation study reported by Reiser & Lin (1999) indicates that, for a data set such as that for the seven pathologists, the Type 1 error rate for X^2 will be more or less the same as the nominal value (but that for G^2 will be higher), but the Type 2 for both statistics will be much inflated (that is, the tests have very little power to detect lack of fit). Another problem is that it is not at all clear what the degrees of freedom should be for a table with so many zeros. So, where now?

One possible solution, and at least it's simple, is to combine cells prior to construction of either of the two chi-squares (and reduce the degrees of freedom, accordingly). But this is not really satisfactory. Another option is to use an alternative goodness-of-fit statistic, such as Read and Cressie's (1988) power divergence statistic, C_λ^2, given by

$$C_\lambda^2 = \frac{2}{\lambda(\lambda + 1)} \sum_{\text{patterns}} \text{Observed} \left[\left(\frac{\text{Observed}}{\text{Expected}} \right)^\lambda - 1 \right] \tag{5.31}$$

in which a value of $\lambda = 2/3$ has been found to lead to a statistic that is less susceptible to the effects of sparseness than either X^2 of G^2. Note, however, that the simulation studies of

Reiser & Lin (1999) indicated that the improvement was rather small. Note that X^2 and G^2 are both special examples of C_λ^2 (for the former $\lambda = 1$ and for the latter $\lambda \to 0$). Several authors (Bartholomew & Knott, 1999, Collins *et al.*, 1993, and Formann, 2003, for example) recommend the use of one of these goodness-of-fit statistics, but to evaluate their sampling distribution using a *Monte Carlo* simulation (the parametric bootstrap). The algorithm for the parametric bootstrap is described by Bartholomew and Knott as follows:

1. Fit the desired model in the usual way.
2. Generate a random sample (of the same size) from the population in which the parameter values are equal to those estimated for the actual sample.
3. Fit the model in each case and compute the chosen test of fit.
4. Compare the actual value of the statistic with the bootstrap sampling distribution.

For an extreme example of sparseness, see Aitkin *et al.* (1981) who used the method on a table with 2^{28} potential cells but with a sample of only 468 subjects. Langeheine *et al.* (1996) illustrate the procedure using a variety of data sets. A nice case study is provided by Goetghebeur *et al.* (2000). Table 5.13 illustrates the results of applying parametric boot-strapping to the latent class model for Holmquist's data (see also Magidson & Vermunt, 2003). Here we present the results of only 50 simulations in order to give the reader a feel for the behaviour of both the X^2 and G^2 statistics. First, note that their mean is around 20, nowhere near 112 (the mean of a χ^2-variate with 112 degrees of freedom). Second, the highest values obtained here are below 40 and, again, are nowhere near as high as the two observed statistics (92.65 and 62.37, respectively). Using 1000 simulations indicates that $P < 0.001$. Our conclusions are that (a) the observed chi-squares should not be compared to the distribution function of a χ^2 with 112 degrees of freedom, and (b) the model clearly does not fit!

Having looked at a global test of fit, and decided that the model is inadequate, Bartholomew and Knott (1999) recommend that one then search for the reasons for the lack of fit. These might include anomalous tests (pathologists), for example. The search is based on the patterns of residuals calculated from marginal frequencies of various orders. The first order marginal frequencies (the observed and predicted total number of biopsies with carcinoma identified by each of the seven pathologists, for example) tell us nothing about the associations between the pathologists' ratings. The second, third and higher order marginal frequencies do. Here, for illustration, we concentrate on the second and third order counts. Bear in mind, however, that although conditional independence of the individual ratings implies conditional independence in the second and third order marginals, the reverse is not necessarily true.

Consider K tests or raters, T_1 to T_K (in our example $K = 7$). The first order margin for test k is simply the sum of the 1s for Test (Rater) k. The observed two-way marginal count (O_{km}) for T_k and T_m is simply the sum of the subjects (biopsies) for which both tests have values of 1. That is,

$$O_{km} = \sum_{\text{subjects}} T_k T_m \tag{5.32}$$

The corresponding expected frequencies (E_{km}) are given by

$$E_{km} = n\left[\pi\alpha_k\alpha_m + (1 - \pi)(1 - \beta_k)(1 - \beta_m)\right] \tag{5.33}$$

Table 5.13 Parametric bootstrap for latent class model for Holquist pathologists' biopsy data – 50 realizations of the chi-square goodness-of-fit statistics.

Observed $X^2 = 92.65$ (112 d.f.):   Observed $G^2 = 62.37$ (112 d.f.).

Replication	X^2	G^2	Replication	X^2	G^2
1	22.738	23.759	26	34.989	24.086
2	22.805	21.981	27	9.204	10.967
3	17.660	14.598	28	19.120	17.330
4	8.668	9.591	29	14.093	11.816
5	15.474	16.689	30	25.787	22.829
6	25.638	24.136	31	17.413	17.356
7	16.659	18.676	32	25.942	26.753
8	34.033	24.986	33	20.976	23.681
9	26.674	20.634	34	8.371	8.858
10	31.517	36.043	35	29.551	23.895
11	24.501	23.759	36	20.260	20.630
12	18.681	18.364	37	22.841	21.024
13	19.900	19.791	38	25.352	24.329
14	35.790	22.754	39	24.819	24.411
15	15.657	15.433	40	22.093	21.55
16	11.293	12.209	41	22.631	17.728
17	17.229	18.373	42	31.036	22.601
18	23.862	20.739	43	12.103	14.249
19	21.763	26.553	44	21.490	24.342
20	22.221	21.782	45	18.745	18.527
21	37.619	38.395	46	19.331	18.327
22	24.063	13.889	47	13.198	14.394
23	37.973	21.517	48	20.938	21.573
24	24.612	26.547	49	13.486	13.376
25	24.363	20.596	50	23.758	18.667

Summary

	X^2	G^2
Mean	21.974	20.302
Variance	51.453	32.986
Percentiles		
1%	8.371	8.858
5%	9.204	10.967
10%	12.651	12.793
25%	17.413	17.330
50%	22.157	20.687
75%	25.352	23.759
90%	32.775	25.767
95%	35.790	26.753
99%	37.973	38.395
Largest	37.973	38.395

where n is the sample size (118 in our present example) and, of course, substituting the estimates of the parameters for their true values. The associated residual (R_{km}) is then defined by Bartholomew and Knott as

$$R_{km} = \frac{\left(O_{km} - E_{km}\right)^2}{E_{km}}$$

(5.34)

Table 5.14 Second order residuals for marginal frequencies for pathologists' biopsy ratings*.

(a) Response pattern $(1, 1)$

	Pathologist					
	A	B	C	D	E	F
B	0.10					
C	0.00	0.00				
D	0.00	0.01	0.11			
E	0.01	0.68	0.02	0.00		
F	0.00	0.01	0.21	**2.21**	0.01	
G	0.00	0.32	0.00	0.00	0.22	0.00

(b) Response pattern $(0, 1)$

	Pathologist					
	A	B	C	D	E	F
B	1.09					
C	0.00	0.00				
D	0.00	0.01	0.13			
E	0.05	**2.38**	0.96	0.15		
F	0.00	0.00	0.15	1.62	0.01	
G	0.00	1.06	–	–	1.11	–

(c) Response pattern $(1, 0)$

	Pathologist					
	A	B	C	D	E	F
B	0.32					
C	–	0.08				
D	–	0.54	0.36			
E	0.03	**4.40**	0.03	0.00		
F	0.00	0.42	0.65	**2.61**	0.53	
G	0.00	**3.61**	0.00	0.00	1.96	0.00

(d) Response pattern $(0, 0)$

	Pathologist					
	A	B	C	D	E	F
B	0.18					
C	0.00	0.00				
D	0.00	0.01	0.04			
E	0.01	1.41	0.02	0.00		
F	0.00	0.01	0.06	0.40	0.01	
G	0.00	0.58	0.00	0.32	0.06	0.00

*Residual $=$ (Observed $-$ Expected)^2/Expected; not defined ($-$) for cells with a zero expected value.

For the present example, these are given in Table 5.14 (from output provided by Bartholomew and Knott's *LATCLASS* program). Note that they should not be confused with the multinomial residuals that contribute to the global Pearson chi-square in Equation (5.30). An inspection of these residuals should reveal where the latent class model may be performing

Table 5.15 Third order residuals for marginal frequencies of pathologists' biopsy ratings (response pattern 1, 1, 1).

ABC	0.00	BCG	0.00
ABD	0.01	BDE	0.00
ABE	0.04	BDF	**2.45**
ABF	0.01	BDG	0.01
ABG	0.01	BEF	0.04
ACD	0.11	BEG	0.36
ACE	0.02	BFG	0.01
ACF	0.21	CDE	0.20
ACG	0.00	CDF	**3.17**
ADE	0.00	CDG	0.11
ADF	**2.21**	CEF	0.31
ADG	0.00	CEG	0.02
AEF	0.01	CFG	0.21
AEG	0.00	DEF	**2.51**
AFG	0.00	DEG	0.00
BCD	0.18	DFG	**2.21**
BCE	0.01	EFG	0.01
BCF	0.28		

* Residual = (Observed − Expected)^2/Expected.

poorly. Table 5.15 provides comparable third order marginal residuals (again produced by *LATCLASS*). Reiser & Lin (1999) and Bartholomew & Leung (2002) have derived global significance tests for second order marginal residuals. One could also use the parametric bootstrap to examine their sampling distributions. Instead, however, we will move on to have a look at observed and expected correlations between tests (pathologists) and the corresponding residuals (simply the difference between the two – see Qu *et al.*, 1996). Letting $\mu_k = \Pr(T_k = 1)$, the correlation for Tests k and m is

$$\frac{\Pr(T_k = 1, T_m = 1) - \mu_k \mu_m}{\sqrt{\mu_k(1 - \mu_k)\mu_m(1 - \mu_m)}} \tag{5.35}$$

For the observed correlation, the estimates for μ_k and $\Pr(T_k = 1, T_m = 1)$ are given by

$$\frac{1}{n} \sum_{subjects} T_k \tag{5.36}$$

and

$$\frac{1}{n} \sum_{subjects} T_k T_m \tag{5.37}$$

The expected values for μ_k and $\Pr(T_k = 1, T_m = 1)$ are

$$\pi \alpha_k + (1 - \pi)(1 - \beta_k) \tag{5.38}$$

and

$$\pi \alpha_k \alpha_m + (1 - \pi)(1 - \beta_k)(1 - \beta_m) \tag{5.39}$$

Table 5.16 Correlations between pathologists' biopsy ratings.

(a) Observed

			Pathologist				
	A	B	C	D	E	F	G
A	–						
B	0.68	–					
C	0.70	0.51	–				
D	0.54	0.43	0.54	–			
E	0.71	0.75	0.64	0.46	–		
F	0.46	0.36	0.49	0.57	0.42	–	
G	0.79	0.76	0.70	0.54	0.81	0.46	–

(b) Expected according to the latent class model

			Pathologist				
	A	B	C	D	E	F	G
A	–						
B	0.59	–					
C	0.70	0.52	–				
D	0.54	0.41	0.48	–			
E	0.69	0.52	0.60	0.47	–		
F	0.46	0.35	0.41	0.31	0.40	–	
G	0.79	0.59	0.70	0.54	0.69	0.46	–

(c) Residual (Observed-Expected)*

			Pathologist				
	A	B	C	D	E	F	G
A	–						
B	0.09	–					
C	0.00	−0.01	–				
D	0.00	0.02	0.06	–			
E	0.02	**0.24**	0.03	−0.01	–		
F	0.00	0.02	0.08	**0.26**	0.02	–	
G	0.00	**0.17**	0.00	0.00	0.13	0.00	–

* Three largest residuals in **bold.**

Again, we substitute estimates of the parameters for their true values. The results are shown in Table 5.16. As before, the sampling distributions of the residuals could be examined using the parametric bootstrap (Qu *et al.*, 1996). A variation on the same theme is to investigate differences between log(odds–ratios) from observed 2×2 margins and those obtained using the corresponding expectations according to the fitted model (Garrett & Zeger, 2000). Anyway, we conclude that there are departures from conditional independence in the biopsy ratings and that these appear to primarily involve Pathologists D and F, but possibly others. We will return to this problem in Section 5.9.

Even though our model (Model (1)) does not appear to fit, we might be tempted to test the effects of introducing some parameter constraints (i.e. comparison of nested models). We have previously remarked on the similarity between Pathologists A, B, E and G, and also between C, D and F. We could try simultaneously introducing the following

Table 5.17 Likelihood-ratio chi-square (G^2) for test of equality of pathologists (sensitivities for all seven the same; specificities for all seven the same).

(a) Standard test results

	X^2	G^2	d.f.
No constraints (1)	92.648	62.366	112
Two groups of pathologists (2)	130.976	91.897	122
All pathologists the same (3)	748.740	269.427	124
Change from (1)–(2)	38.328	29.531	10
Change from (1)–(3)	656.092	207.062	12
Change from (2)–(3)	617.764	177.530	2

(b) Parametric bootstrap (1000 simulations) for comparison of (1) and (3) – Distribution function for G^2 and a χ^2_{12}-variate (for comparison)

	G^2	χ^2_{12}
1%	3.53	3.57
5%	5.56	5.23
10%	6.41	6.30
25%	8.56	8.44
50%	11.74	11.34
75%	15.91	14.85
90%	19.51	18.55
95%	22.00	21.03
99%	28.40	26.22

constraints: $\alpha_A = \alpha_B = \alpha_E = \alpha_G, \alpha_C = \alpha_D = \alpha_F, \beta_A = \beta_B = \beta_E = \beta_G$, and $\beta_C = \beta_D = \beta_F$ (Model (2)). We could also simultaneously constrain all seven sensitivities (the αs) to be the same and all specificities (the βs) to be equal too (Model (3)). The results of doing this in *Mplus* (see Appendix 5(b) for setup of Model (2)) are shown in Table 5.17(a). At face value it looks as if we should not be introducing any of these constraints. The changes in both X^2 and G^2 are very high and if they were distributed as χ^2 then they would be very highly significant. Note, however, how markedly different the two forms of chi-square are. So, knowing the problems arising from having a very sparse table of data, are these tests safe? Table 5.17(b) shows the results of a *Monte Carlo* bootstrap simulation for the comparison of Models (1) and (3). This was carried out using *Mplus*. First, we simulated 1000 sets of results using Model (3) to generate the data (with the parameter estimates obtained from the fitting of Model (3) to the Holmquist data: $\alpha = 0.766$ and $\beta = 0.929$) but an unconstrained model (Model (1)) for the analysis. From these we obtained values for the log likelihood. Then we repeated the creation of exactly the same sets of data (i.e. using the same seed and model to generate the data) and analysed them using Model (3) to obtain new log likelihood values. Twice the difference between the two provided the required G^2 statistics. The percentiles for these are given in Table 5.17 along with the corresponding values for the χ^2_{12} distribution. They look remarkably similar. So G^2, at least, does appear to be approximating its asymptotic distribution in this particular example. This is in accord with the simulation studies of Agresti & Yang (1987) but should, however, should be compared with those of Goetghebeur *et al.* (2000) who found marked disparities in their analysis of tests for visceral leishmaniasis. More work needs to be done. At the end of their paper on bootstrapping goodness-of-fit measures, Langeheine *et al.* (1996) comment:

> Another extension of our work would be to bootstrap likelihood ratio tests of nested models. Then the often-heard assertion that asymptotically valid tests can be used there even with moderately sparse data could be confronted with bootstrap tests. (p 513)

So far, we have a slightly mixed set of results in the case of latent class analysis of sparse contingency tables, but investigators should be warned not to automatically trust tests based on their asymptotic properties.

Perhaps before leaving this section we should point out that despite difficulties of drawing valid statistical inferences from sparse tables, they are not necessarily the result of poor design. Clearly the ideal would be to obtain a sample large enough for the problem to go away. But the cost in many cases would be prohibitive. So we have to learn to carry out an effective analysis with the data we've got at hand. Goetghebeur *et al.* (2000) suggests looking at the results of different subsets of tests to check the sensitivity of the results and contribute to safer inferences.

5.8 The implications of conditional (local) dependencies

We have found that in the case of the Holmquist biopsy data, a simple latent class model, with assumed conditional independence of the ratings by each of the seven pathologists, does not fit the data. We can search for better fitting models (which we will describe later in this section) but first it might be instructive to check the sensitivity of these results to deletion of one or more of the pathologists' ratings (see, for example, Magidson & Vermunt, 2003). We start by deleting each of the seven pathologists in turn. Bootstrap *P*-values produced from 500 simulations by the package *Latent GOLD* (Magidson & Vermunt, 2000) are given in Table 5.18(a). Still the models do not fit. The G^2 goodness-of-fit statistics and their associated *P*-values are also presented to emphasize once again their lack of validity in this situation. We now try deleting each pair of pathologists in turn (21 combinations). The results of this exercise are given in Table 5.18(b). All but one of the resulting models (that involving deletion of pathologists *B* and *E*) still fail to fit the data. The parameter estimates derived from fitting the simple latent class model for pathologists *A*,*C*,*D*,*F* and *G* (i.e. excluding *B* and *E*) are given in Table 5.19 (2nd column – the first column shows the results from the original model for comparison). The surprising conclusion is how little they have changed in moving from the invalid model to one that fits.

How might we parameterize our modified latent class models to allow for lack of conditional independence? Before we go further, we might consider an alternative parameterization of a latent class model (with conditional independence) as log-linear model for the expected cell counts. (Haberman, 1979). This allows for main effects of disease (the latent class *D*) and disease indicators (pathologist ratings, for example) and two-way associations (interactions) between disease and each of the indicators, and nothing else:

$$\log m_{hijk} = \lambda + \lambda_h^A + \lambda_i^B + \lambda_j^C + \lambda_k^D + \lambda_{hk}^{AD} + \lambda_{ik}^{BD} + \lambda_{jk}^{CD} \qquad (5.40)$$

Of course, we never observe the disease status *D* so that, in practice, using the standard log-linear modelling notation, and using three ratings (*A*, *B* and *C* with levels $h = 0, 1, i = 0, 1$

Table 5.18 Goodness-of-fit for subsets of pathologists' biopsy ratings.

(a) Subsets of six (50 d.f.)

Selected ratings	G^2	P-value	Bootstrap P-value
BCDEFG	42.81	0.760	<0.001
ACDEFG	31.26	0.980	0.006
ABDEFG	42.62	0.760	0.004
ABCEFG	46.72	0.600	<0.001
ABCDFG	35.92	0.930	<0.001
ABCDEG	48.04	0.550	<0.001
ABCDEF	42.32	0.770	0.002

(b) Subsets of five (20 d.f.)

Selected ratings	G2	P-value	Bootstrap P-value
CDEFG	24.23	0.230	0.004
BDEFG	28.53	0.097	<0.001
BCEFG	24.68	0.210	<0.001
BCDFG	25.43	0.180	0.004
BCDEG	25.94	0.170	0.004
BGDEF	33.65	0.029	<0.001
ADEFG	24.90	0.210	0.002
ACEFG	17.33	0.630	0.018
ACDFG	**11.92**	**0.920**	**0.144**
ACDEG	20.06	0.450	0.012
ACDEF	16.48	0.690	0.048
ABEFG	21.92	0.340	0.028
ABDFG	29.23	0.083	<0.001
ABDEG	23.26	0.280	0.020
ABDEF	29.99	0.070	<0.001
ABCFG	25.01	0.200	<0.001
ABCEG	35.69	0.017	<0.001
ABCEF	26.02	0.160	<0.001
ABCDG	25.05	0.200	0.002
ABCDF	18.90	0.530	0.018
ABCDE	28.97	0.088	<0.001

and $j = 0, 1$, respectively) we are fitting models to observations of the following structure (see Agresti, 1992, for example):

$$\log m_{hij} = \lambda + \lambda_h^A + \lambda_i^B + \lambda_j^C + \log\left[\sum_k \exp\left(\lambda_k^D + \lambda_{hk}^{AD} + \lambda_{ik}^{BD} + \lambda_{jk}^{CD}\right)\right] \quad (5.41)$$

Here, m_{hij+} is the expected count after adding over the two levels ($k = 0, 1$) of the latent class indicator, D. The terms inside the square brackets are also being summed over the levels of D. The parameters are constrained so that

$$\sum_h \lambda_h^A = \sum_i \lambda_i^B = \sum_j \lambda_j^C = \sum_k \lambda_k^D = 0$$

and

$$\sum_h \lambda_{hk}^{AD} = \sum_i \lambda_{ik}^{BD} = \sum_j \lambda_{jk}^{CD} = 0$$

Table 5.19 Parameter estimates for modified latent models for the Holmquist biopsy data.

Parameter	Original model Estimate (s.e.)	After dropping Pathologists B & E Estimate (s.e.)	Model with direct effects Estimate (s.e.)*
α_A	1.000 (0.000)	1.000 (0.000)	1.00 (0.00)
α_B	0.983 (0.017)	–	0.98 (0.02)
α_C	0.761 (0.057)	0.774 (0.061)	1.00 (0.00)
α_D	0.541 (0.065)	0.551 (0.069)	0.58 (0.07)
α_E	0.979 (0.021)	–	1.00 (0.00)
α_F	0.423 (0.065)	0.430 (0.068)	0.47 (0.07)
α_G	1.000 (0.000)	1.000 (0.000)	1.00 (0.00)
β_A	0.883 (0.043)	0.868 (0.053)	0.72 (0.06)
β_B	0.646 (0.064)	–	0.53 (0.06)
β_C	1.000 (0.000)	1.000 (0.000)	1.00 (0.00)
β_D	1.000 (0.000)	1.000 (0.000)	0.92 (0.03)
β_E	0.777 (0.055)	–	0.65 (0.06)
β_F	1.000 (0.000)	1.000 (0.000)	0.95 (0.02)
β_G	0.883 (0.117)	0.868 (0.053)	0.72 (0.06)
$\mathrm{logit}(1 - \pi)$	-0.005 (0.185)	0.030 (0.194)	**
b_0	–	–	2.212 (0.572)
b_1	–	–	0.607 (0.242)

* From the analysis of Qu *et al.*, 1996.
** Not known.

In terms of this model parameterization, the sensitivity of rater A, for example, is given by

$$
\Pr(A = 1 \mid D = 1, B = i, C = j)
$$
$$
= \frac{\exp\!\left(\lambda + \lambda_1^A + \lambda_i^B + \lambda_j^C + \lambda_1^D + \lambda_{11}^{AD} + \lambda_{i1}^{BD} + \lambda_{j1}^{CD}\right)}{\left[\begin{array}{l}\exp\!\left(\lambda + \lambda_0^A + \lambda_i^B + \lambda_j^C + \lambda_1^D + \lambda_{01}^{AD} + \lambda_{i1}^{BD} + \lambda_{j1}^{CD}\right) \\ + \exp\!\left(\lambda + \lambda_1^A + \lambda_i^B + \lambda_j^C + \lambda_1^D + \lambda_{11}^{AD} + \lambda_{i1}^{BD} + \lambda_{j1}^{CD}\right)\end{array}\right]}
$$
$$
= \frac{\exp\!\left(\lambda + \lambda_1^D\right)\exp\!\left(\lambda_i^B + \lambda_j^C + \lambda_{i1}^{BD} + \lambda_{j1}^{CD}\right)\exp\!\left(\lambda_1^A + \lambda_{11}^{AD}\right)}{\left[\begin{array}{l}\exp\!\left(\lambda + \lambda_1^D\right)\exp\!\left(\lambda_i^B + \lambda_j^C + \lambda_{i1}^{BD} + \lambda_{j1}^{CD}\right)\exp\!\left(\lambda_0^A + \lambda_{01}^{AD}\right) \\ + \exp\!\left(\lambda + \lambda_1^D\right)\exp\!\left(\lambda_i^B + \lambda_j^C + \lambda_{i1}^{BD} + \lambda_{j1}^{CD}\right)\exp\!\left(\lambda_1^A + \lambda_{11}^{AD}\right)\end{array}\right]}
$$
$$
= \frac{\exp\!\left(\lambda_1^A + \lambda_{11}^{AD}\right)}{\exp\!\left(\lambda_0^A + \lambda_{01}^{AD}\right) + \exp\!\left(\lambda_1^A + \lambda_{11}^{AD}\right)} \tag{5.42}
$$

Equivalently,

$$
\frac{\Pr\!\left(A = 1 \mid D = 1, B = i, C = j\right)}{\Pr\!\left(A = 0 \mid D = 1, B = i, C = j\right)} = \frac{\exp\!\left(\lambda_1^A + \lambda_{11}^{AD}\right)}{\exp\!\left(\lambda_0^A + \lambda_{01}^{AD}\right)} \tag{5.43}
$$

A similar expression is easily derived for the specificity of A. Now, if we have evidence of departure of conditional independence (of the ratings A and B, say) we can add a further interaction term to Equation (5.40), λ_{hi}^{AB}, which measures the 'direct effect' between A

and B. We could let the direct effect of A and B depend on the value of D through the introduction of a three-way interaction (Espeland & Handelman, 1989; Yang & Becker, 1997; Magidson & Vermunt, 2003). Returning to the Holmquist biopsy data, an exploratory analysis using *Latent GOLD* indicates that we need to include three direct effects to the latent class model in order to get an adequate fit (Bootstrap P-value = 0.372). These are between B and E, between F and D, and between B and G. The problem, however, is that the parameter estimates can now no longer have a simple direct interpretation in terms of sensitivity and specificity of the raters (see Zhou *et al.*, 2002). Interpretation is not obvious. Returning to the simpler example with three ratings (A, B and C) and allowing for a direct effect between A and B, we now have

$$\Pr\big(A = 1 \mid D = 1,\, B = i,\, C = j\big) = \frac{\exp\big(\lambda_1^A + \lambda_{11}^{AD} + \lambda_{1i}^{AB}\big)}{\exp\big(\lambda_0^A + \lambda_{01}^{AD} + \lambda_{0i}^{AB}\big) + \exp\big(\lambda_1^A + \lambda_{11}^{AD} + \lambda_{1i}^{AB}\big)}$$

(5.44)

or

$$\frac{\Pr\big(A = 1 \mid D = 1,\, B = i,\, C = j\big)}{\Pr\big(A = 0 \mid D = 1,\, B = i,\, C = j\big)} = \frac{\exp\big(\lambda_1^A + \lambda_{11}^{AD} + \lambda_{1i}^{AB}\big)}{\exp\big(\lambda_0^A + \lambda_{01}^{AD} + \lambda_{0i}^{AB}\big)}$$

(5.45)

which has the added λ_{hi}^{AB} terms and, therefore, is dependent on the value of B. Yang and Becker (1997) have provided a solution to this problem by fitting a marginal latent class model. Their models involve specifying the univariate margins of the raters or tests (A, B and C, for example) in terms of logits. For example

$$\log\left(\frac{\Pr\big(A = 1 \mid D = k\big)}{\Pr\big(A = 0 \mid D = k\big)}\right) = \alpha_k^A$$

(5.46)

The log–odds ratio for the association between A and B is defined as

$$\log\left(\frac{\Pr\big(A = 0\ \&\ B = 0 \mid D = k\big)\Pr\big(A = 1\ \&\ B = 1 \mid D = k\big)}{\Pr\big(A = 0\ \&\ B = 1 \mid D = k\big)\Pr\big(A = 1\ \&\ B = 0 \mid D = k\big)}\right) = \psi_k^{AB}$$

(5.47)

For our example involving a direct effect between A and B, we fit the above model to estimate α_k^A, $\alpha_k^B \alpha_k^C$ and ψ_κ^{AB} ($k = 0, 1$) with all other two-way interactions and all three-way interactions constrained to be zero. The required sensitivity and specificity for A, for example, are then given by

$$\text{Sensitivity} = \frac{\exp\big(\alpha_1^A\big)}{1 + \exp\big(\alpha_1^A\big)}$$

(5.48)

and

$$\text{Specificity} = \frac{1}{1 + \exp\big(\alpha_0^A\big)}$$

(5.49)

respectively. Yang and Becker (1997) used a purpose-written EM algorithm to obtain maximum likelihood estimates and likelihood ratio statistics to test goodness-of-fit. Further technical details, together with illustrative examples, can be found in Yang and Becker (1997), Becker and Yang (1998) and in Zhou *et al.* (2002).

Yang and Becker's marginal latent class model is a possibility for the Holmquist biopsy data but, instead of pursuing this further, we will here consider another option, proposed by Qu *et al.* (1996). Returning to the standard latent class model (as traditional parameterized – i.e. not the log-linear model as above) with conditional independence, Qu *et al.* postulate the existence of an additional (continuous) latent variable, t, to explain the conditional dependencies among the raters or tests. The variable t is assumed to be distributed according to the standard normal distribution. The probability of a positive result from rater (test) k, given true status, D, is a monotonic function of t. Qu *et al.* (1996), describe this relationship using the probit distribution:

$$\Pr(T_k = 1 \,|\, D, t) = \Phi(a_{kD} + b_{kD}t) \tag{5.50}$$

Not that for K tests this involves the introduction of K new parameters. If all of the b_{kD} are equal to zero, then we are back to conditional independence. Other special cases include an equality constraint on all of the b_{kD} given the disease status, D, and setting several (but not all) of their values to zero. The latter constraints allow the modelling of the direct effects between tests in terms of the random variable, t. Note that the sensitivity and specificity of Test k are given by

$$\Pr(T_k = 1 \,|\, D = 1) = \int_{-\infty}^{\infty} \Phi(a_{k1} + b_{k1}t)\, \phi(t)dt$$

$$= \Phi\left\{ \frac{a_{k1}}{(1 + b_{k1}^2)^{1/2}} \right\} \tag{5.51}$$

and

$$\Pr(T_k = 0 \,|\, D = 0) = 1 - \int_{-\infty}^{\infty} \Phi(a_{k0} + b_{k0}t)\, \phi(t)dt$$

$$= 1 - \Phi\left\{ \frac{a_{k1}}{(1 + b_{k1}^2)^{1/2}} \right\} \tag{5.52}$$

The probability function for the K test results on a given subject, conditional on true status, D, (i.e. the subject's contribution to the likelihood function) is given by

$$\int_{-\infty}^{\infty} \prod_{k=1}^{K} \Phi(a_{k0} + b_{k0}t)^{T_k} \{1 - \Phi a_{kD} + b_{kD}t\}^{1-T_k} \phi(t)dt \tag{5.53}$$

where $T_k = 0, 1$. The integration in (5.53) is achieved numerically by Gauss-Hermite quadrature and maximization of the likelihood is carried out using an EM algorithm (see Qu *et al.*, 1996, and Qu & Hadgu, 1998, for technical details). Detailed examples of the use of these random effects models are provided by Hadgu and Qu (1998) and Goetghebeur *et al.* (2000). Qu *et al.* used our familiar Holmquist biopsy data for one of their illustrative examples (constraining the b_{kD}s to be the same for all k). Unfortunately their goodness-of-fit assessments were based on chi-square values and these, in turn, seem to have been subject to printing errors. So, it is difficult to assess how well their model actually explained the data, but their parameter estimates are given in the third column of Table 5.19. Note that the estimated value of b_1 is barely significantly different from 0 and it might be dropped from the model. Its effect on the sensitivities is to perhaps increase them very slightly (in the cases where this is possible) but the differences are not marked. The estimate of b_0, however, appears to be highly statistically significant, and its influence is to lower the estimates of the pathologists' specificities.

We have seen a variety of approaches to cope with and model the effects of local dependencies between raters or tests, and each of the approaches can produce models that fit the data and can provide a parsimonious description of the data. Marginal latent class models (Yang & Becker, 1997) and latent class models with random effects (Qu *et al.*, 1996) seem particularly promising. But are they credible as a model of the actual measurement process? It is hard to tell. To date the development of the models appears to be motivated more by mathematics than science. There has been too little stress laid on the design of method comparison studies involving binary assessments, and in particular using replicate assessments (with a few exceptions, such as Baker *et al.*, 1991). One important question to ask is whether disease status is truly binary. Might it be better represented by a quantitative latent trait? Latent trait models for binary data are beyond the scope of the present text and we refer interested readers to Bartholomew and Knott (1999) for further details. The latter authors (pp 135–7) point out that the two families of model are often difficult to distinguish, even to the extent of providing similar parameter estimates. Of particular interest in this context is a model for rater agreement data described by Uebersax and Grove (1993). This involves specifying, say, two latent classes (diseased or well, for example) but within each class the observed binary indicators (tests or ratings) are reflecting an underlying quantitative latent trait.

5.9 Use of instrumental variables

We now move from one extreme to the other. Let's consider a situation in which we have used only two tests or raters. Our proposed latent class model is under-identified. Hui and Walter (1980) considered the performance of two diagnostic tests for tuberculosis (the Mantoux test and the Tine test). Gart and Buck (1966) wished to compare two tests for the venereal disease, syphilis – the Venereal Disease Research Laboratory Slide Flocculation test (VDRL) and the Flourescent Treponeal Antibody test (FTA). Neither pair of authors had access to a definitive diagnosis. The approaches taken by these authors are different, but complementary. Hui and Walter (1980) introduced what, in effect, is an instrumental variable. They had data collected from two populations in which they were confident that the prevalence of tuberculosis differed considerably. So disease status, D, was associated with Population but, conditional on D, the test results (Mantoux and Tine) were assumed to be independent of each other and of Population. This enabled them to estimate the sensitivities and specificities of the two tests as required. The binary variable, Population, can be used as a third manifest indicator of disease and the data fitted using the standard latent class model for three observed binary indicators (see also, Walter, 1984). Gart and Buck

Table 5.20 Frequencies of syphilis test results according to age group.

Age group	FTA	VDRL	Count
5–14 (=1)	1	1	1
	1	0	4
	0	1	10
	0	0	62
15–24 (=2)	1	1	5
	1	0	2
	0	1	5
	0	0	31
25–34 (=3)	1	1	14
	1	0	6
	0	1	14
	0	0	27
35–44 (=4)	1	1	20
	1	0	5
	0	1	17
	0	0	19
45+ (=5)	1	1	18
	1	0	5
	0	1	9
	0	0	17

Source: Gart & Buck, 1966.

(1966) assumed that the sensitivity and specificity of the FTA test was known (from past research) and then fitted the standard latent class model, which was now just identified. We will illustrate both approaches using data provided by Gart and Buck (1966). They are not mutually exclusive, provided we have sufficient information, and the Gart and Buck data enable us to check the sensitivity of the results to the assumptions made. We assume conditional independence of the two tests and the instrumental variable in all models.

Table 5.20 provides syphilis test results (that is, FTA and VDRL) for five different age groups (5–14, 15–24, 25–34, 35–44 and 44+). The prevalence of syphilis appears to be strongly dependent on age. Gart and Buck (1966) assumed that the sensitivity and specificity of the FTA test (α_F and β_F, say) did not change with age and were known from prior work to be 0.95 and 0.90, respectively. They then used these values to provide estimates of the sensitivity and specificity of the VDRL test (i.e. α_V and β_V). Using the notation in Table 5.2, it can be shown (Gart & Buck, 1966; Zhou *et al.*, 2002) that the maximum likelihood estimates of α_V and β_V are given by

$$\hat{\alpha}_V = \frac{\beta_F(n_{11} - n_{01}) - n_{01}}{n\beta_F - (n_{01} + n_{00})} \tag{5.54}$$

and

$$\hat{\beta}_V = \frac{\alpha_F(n_{10} + n_{00}) - n_{00}}{n\alpha_F - (n_{11} + n_{10})} \tag{5.55}$$

They can also be found using a numerical maximum likelihood algorithm for a latent class model (in *Mplus*, for example) by simply constraining α_A and β_A to have values of 0.95 and 0.90, respectively. The results are given in Table 5.21(a). In the first two age groups there

Table 5.21 Estimates from Gart and Buck syphilis test data.

(a) Analysis of separate age groups

Age group	Estimates (s.e.)				
	$\text{logit}(1 - \pi)$	α_F	β_F	α_V	β_V
5–14	15.691 (9.531)	0.95*	0.90*	1.000 (0.000)	0.857 (0.040)
15–24	2.096 (0.542)	0.95*	0.90*	1.000 (0.000)	0.862 (0.060)
25–34	1.004 (0.360)	0.95*	0.90*	0.806 (0.140)	0.668 (0.076)
35–44	0.556 (0.320)	0.95*	0.90*	0.862 (0.098)	0.540 (0.086)
45+	0.263 (0.341)	0.95*	0.90*	0.845 (0.101)	0.675 (0.098)

(b) Using instrumental variable methods (using age group coding as a quantitative covariate)

	α_F	β_F	α_V	β_V
Model for $\text{logit}(\pi)$	0.95*	0.90*	0.883 (0.060)	0.754 (0.031)
	$\gamma_0 = -3.873$ (s.e. 0.464); $\gamma_1 = 0.797$ (s.e. 0.124)			
	α_F	β_F	α_V	β_V
Model for $\text{logit}(\pi)$	0.595 (0.083)	0.959 (0.029)	0.768 (0.079)	0.889 (0.046)
	$\gamma_0 = -3.169$ (s.e. 0.527); $\gamma_1 = 0.949$ (s.e. 0.188)			

(c) Using instrumental variable methods (using age group as a qualitative factor)

	α_F	β_F	α_V	β_V
Model for $\text{logit}(\pi)$	0.95*	0.90*	0.879 (0.061)	0.754 (0.031)
	$\gamma_1 = -19.293$ (0.447)			
	$\gamma_2 = 17.125$ (0.698); $\gamma_3 = 18.264$ (0.540);			
	$\gamma_4 = 18.828$ (0.569); $\gamma_5 = 19.034$ (0.569)			
	α_F	β_F	α_V	β_V
Model for $\text{logit}(\pi)$	0.625 (0.064)	0.945 (0.027)	0.809 (0.061)	0.876 (0.038)
	$\gamma_1 = -4.252$ (2.233)			
	$\gamma_2 = 2.752$ (2.221); $\gamma_3 = 4.233$ (2.168);			
	$\gamma_4 = 4.996$ (2.186); $\gamma_5 = 4.861$ (2.203)			

* Constraint.

are boundary problems so the estimates provided by *Mplus* are not exactly the same as those obtained via (5.54) and (5.55). The rest are.

We can now add in a part of the overall model that predicts the unknown prevalence (π_i) of the disease in age group i. If we treat age group as a quantitative measurement we can fit a logistic model of the following form:

$$\log\left(\frac{\pi_i}{1 - \pi_i}\right) = \gamma_0 + \gamma_1 \, \text{Age} \tag{5.56}$$

or

$$\pi_i = \frac{\exp(\gamma_0 + \gamma_1 \text{Age})}{1 + \exp(\gamma_0 + \gamma_1 \text{Age})} \tag{5.57}$$

where *Age* takes values from 1 to 5 (see Nagelkerke *et al.*, 1988). Alternatively, we could fit the effect of *Age* through four binary (0/1) dummy variables A_2 (1 for age group 2, 0 otherwise),

A_3 (1 for age group 3, 0 otherwise), A_4 (1 for age group 4, 0 otherwise) and A_5 (1 for age group 5, 0 otherwise), so that

$$\log\left(\frac{\pi_i}{1 - \pi_i}\right) = \gamma_i + \sum_{j=2}^{5} \gamma_j A_j \tag{5.58}$$

or

$$\pi_i = \frac{\exp\left(\gamma_i + \sum_{j=2}^{5} \gamma_j A_j\right)}{1 + \exp\left(\gamma_i + \sum_{j=2}^{5} \gamma_j A_j\right)} \tag{5.59}$$

This is the approach taken by Hui and Walter (1980) but in their example there are only two groups. The joint likelihood for all five age groups is now given by

$$L = \prod_{i=1}^{5} \left[\left\{\pi_i(1 - \alpha_F)(1 - \alpha_V) + (1 - \pi_i)\beta_F\beta_V\right\}^{n_{i11}} \left\{\pi_i(1 - \alpha_F)\alpha_V \right. \right.$$
$$+ (1 - \pi_i)\beta_F(1 - \beta_V)\right\}^{n_{i01}} \times \left\{\pi_i\alpha_F(1 - \alpha_V) + (1 - \pi_i)(1 - \beta_F)\beta_V\right\}^{n_{i10}}$$
$$\left.\left\{\pi_i\alpha_F\alpha_V + (1 - \pi_i)(1 - \beta_F)(1 - \beta_V)\right\}^{n_{i00}} \right] \tag{5.60}$$

where π_i can be substituted either from Equation (5.58) or (5.59). The notation for the cell counts should be clear from the context. Now, in addition to the two models for prevalence we can also choose to constrain the values of α_F and β_F (0.05 and 0.90, respectively), as in Gart and Buck's treatment, or leave them to be estimated (the *Mplus* set up for the latter is given in Appendix 5(c)). The results are given in Tables 5.21(b) and 5.21(c). Note that they are not entirely consistent. Treating age group as a qualitative factor appears to be better than fitting a linear relationship on the logistic scale (the standard errors of the parameter estimates are smaller). Letting α_F and β_F be free to be estimated produces lower estimates for these two parameters than those assumed to hold from prior evidence (particularly in the case of α_F), and also to slightly different estimates of α_V and β_V. Assuming that we know the values of the sensitivity and specificity of the FTA test, however, enables us to get more precise estimates of the characteristics for the VDRL test. We cannot tell which results are the correct ones. Perhaps we should not even expect to get at the truth. But the changes in the estimates as we change our assumptions should prompt us to think more carefully about the processes generating the data. This is the great strength of modelling, not that it necessarily produces the right answer.

A further use for instrumental variables in latent class models is to help check the conditional independence assumptions when we only have ratings using three methods (Nagelkerke *et al.*, 1988). Under normal circumstances we have no degrees of freedom with which to test the latent class model. If we can find a convincing instrumental variable (that is, one which is strongly related to prevalence but conditionally independent of the three test results) then this gives us a handle on the problem. Nagelkerke *et al.* illustrate the ideas using three tests to detect a prescribed drug for the treatment of leprosy (a Spot test, an ELIZA test and a haemagglutination test) together with the respondents' educational status (no primary schooling, incomplete primary schooling and complete primary schooling). Unfortunately, schooling was not a strong enough predictor of prescribed drug compliance to be an entirely convincing instrumental variable. In principle, however, this is a good idea and more consideration should be given to the collection of data on potential instrumental variables (and, indeed, other covariates) as part of the effective design of method comparison studies.

5.10 Data missing by design

Two-phase designs for the comparison of quantitative measurements methods were discussed in considerable detail in Section 4.13. These designs are also a familiar feature in work on the evaluation of binary diagnostic tests (see, for example, Tosteson *et al.*, 1994; Chock *et al.*, 1997; Walter, 1999; Sullivan Pepe & Alonzo, 2001; Berry *et al.*, 2002; Zhou *et al.*, 2002; Sullivan Pepe, 2003). They have also been suggested for rater agreement studies, involving the estimation of kappa coefficients (Jannarone et al., 1987; Kraemer & Bloch, 1990). Typically a first-phase sample is assessed using one or more fallible diagnostic or screening tests and verification (using supposed infallible 'gold standard' diagnoses, or further fallible tests, including the possibility of replication of those used in phase 1) is carried out on a randomly selected sub-sample of the respondents in phase 2. The selection of the phase 2 participants might be completely at random (see, for example, Sinclair and Gastwith, 1996), or after stratification according to the first phase results and possibly other covariates (Tosteson *et al.*, 1994; Zhou, 1998a). In the case of a relatively rare disease, one might decide to carry all those with evidence of illness forward into the second phase, but sample only a relatively small proportion of those thought to be well according to the first phase tests. In diseases such as cancer, for example, the second phase is likely to involve invasive procedures such as tissue biopsies and there are likely to be strong ethical objections to carrying out these procedures on participants who, so far, have shown no signs of pathology. In this example the second phase sampling fractions for those screened negative in phase 1 would be zero (Walter, 1999; Chock *et al.*, 1997; Sullivan Pepe & Alonzo, 2001; Berry *et al.*, 2002).

The purpose of the material in this section is to simply illustrate how we can fit latent class models for diagnostic test data arising from two phase designs. We know that, unless there is non-participation in phase 2 for reasons other than those determined by the investigators, the missing observations at phase 2 are missing by design (that is, missing at random – MAR). It is quite straightforward to fit latent class models with data assumed to be missing at random and we compare the results with those obtained through the use of a naïve complete case analysis (that is, restricting the analysis to only those participants who provided data in both phases of the study).

The data we will use to illustrate the ideas was collected by S.L. Handelman and analysed by Espeland and Handelman (1989). The data as collected is shown in Table 5.22. Each of five dentists independently assessed 3869 X-rays of molar or pre-molar teeth and decided whether they were sound (coded as 1) or as having caries (coded as 0). There would appear to be no ethical reasons for criticizing the design, but we might ask if we could have obtained essentially the same information if there had been an element of two-phase sampling in the collection of the data. There are 1880 teeth that were given a diagnosis of caries by all five dentists. Might a more efficient design have been possible? Might a full assessment of these apparently healthy teeth by all five dentists have been avoided? With this in mind we will

Table 5.22 Radiographic assessments of 3869 teeth by five dentists. The teeth are classified as being sound (=0) or as having caries (=1).

A	B	C	D	E	Frequency
0	0	0	0	0	1880
0	0	0	0	1	789
0	0	0	1	0	43
0	0	0	1	1	75
0	0	1	0	0	23
0	0	1	0	1	63
0	0	1	1	0	8
0	0	1	1	1	22
0	1	0	0	0	188
0	1	0	0	1	191
0	1	0	1	0	17
0	1	0	1	1	67
0	1	1	0	0	15
0	1	1	0	1	85
0	1	1	1	0	8
0	1	1	1	1	56
1	0	0	0	0	22
1	0	0	0	1	26
1	0	0	1	0	6
1	0	0	1	1	14
1	0	1	0	0	1
1	0	1	0	1	20
1	0	1	1	0	2
1	0	1	1	1	17
1	1	0	0	0	2
1	1	0	0	1	20
1	1	0	1	0	6
1	1	0	1	1	27
1	1	1	0	0	3
1	1	1	0	1	72
1	1	1	1	0	1
1	1	1	1	1	100

Proportion thought to have caries

0.088	0.222	0.128	0.121	0.425

Source: Espeland & Handelman (1989).

select data from the full set in Table 5.22 using three different sampling plans and then go on to analyse the results using a latent class model (either assuming MAR or using a complete case analysis). We know from the work of Espeland and Handelman (1989) and from others (Qu *et al.*, 1996, for example) that the assumption of conditional independence is not tenable for these particular data. Our main purpose, however, is to illustrate the effects of two-phase sampling designs and we will assume that the standard latent class model provides a good enough representation of the data patterns. Clearly the methods could be generalized to the more complex models used by Espeland and Handelman (1989) or by Qu *et al.* (1996).

In our hypothetical two-phase designs we assume that all of the 3869 teeth are assessed by Dentists A and E (these two seem to be producing widely diverging assessments according to the proportions of the teeth they consider to be displaying evidence of caries – see bottom of Table 5.22). For phase 2 (which involves assessments by the three remaining dentists – B, C and D) we adopt three different sampling procedures. In all three procedures we select every tooth for further assessment for which there appears to be some evidence of caries (Dentists A and E both say caries present, or at least one of them does). The

Table 5.23 Results of fitting latent class models to Handelman dental data.

Parameter	Estimates (s.e.)						
	Full data	Plan A MAR	Plan A CC	Plan B MAR	Plan B CC	Plan C MAR	Plan C CC
α_A	0.403 (0.025)	0.531 (0.032)	0.556 (0.031)	0.497 (0.030)	0.507 (0.030)	0.484 (0.030)	0.489 (0.030)
α_B	0.713 (0.023)	0.795 (0.024)	0.734 (0.024)	0.780 (0.025)	0.800 (0.025)	0.771 (0.025)	0.786 (0.026)
α_C	0.598 (0.028)	0.709 (0.031)	0.648 (0.027)	0.680 (0.030)	0.698 (0.031)	0.674 (0.030)	0.692 (0.030)
α_D	0.489 (0.022)	0.540 (0.026)	0.514 (0.025)	0.532 (0.025)	0.545 (0.026)	0.523 (0.026)	0.530 (0.026)
α_E	0.915 (0.085)	0.960 (0.014)	0.927 (0.011)	0.950 (0.015)	0.961 (0.010)	0.947 (0.015)	0.959 (0.010)
β_A	0.989 (0.002)	0.984 (0.003)	0.988 (0.007)	0.986 (0.003)	0.957 (0.007)	0.986 (0.003)	0.963 (0.006)
β_B	0.898 (0.007)	0.819 (0.016)	0.819 (0.016)	0.843 (0.013)	0.842 (0.014)	0.849 (0.012)	0.849 (0.012)
β_C	0.986 (0.003)	0.942 (0.011)	0.944 (0.011)	0.955 (0.009)	0.953 (0.009)	0.964 (0.008)	0.965 (0.009)
β_D	0.968 (0.005)	0.907 (0.012)	0.917 (0.011)	0.929 (0.010)	0.927 (0.010)	0.933 (0.009)	0.932 (0.009)
β_E	0.695 (0.010)	0.662 (0.009)	0.000 (0.000)	0.669 (0.010)	0.177 (0.011)	0.672 (0.010)	0.276 (0.013)
$\mathrm{logit}(\pi)$	−1.411 (0.067)	−1.820 (0.077)	−0.628 (0.085)	−1.720 (0.075)	−0.890 (0.085)	−1.682 (0.076)	−0.952 (0.084)
π	0.20	0.14	0.34	0.15	0.29	0.16	0.2

procedures differ in the way we treat those teeth for which there is no initial evidence of disease (that is, Dentists A and E both say that the tooth is sound). In Plan A we select none of the teeth assessed to be sound by both A and E, in Plan B we randomly select 10% of the teeth assessed to be sound by both A and E, and in Plan C we randomly select 20%. Given the potential number of sound teeth involved, all three sampling plans could have saved quite a lot of time, effort and money. But what of the results? These are shown in Table 5.23.

The striking thing about the results in Table 5.23 is how little is lost by dropping so many observations. With the appropriate method of analysis (assuming that observations are MAR) even the results from the two-phase Plan A (collecting no further assessments on teeth judged healthy by both Dentist A and Dentist E) produce pretty good results. Neither the estimates of the sensitivities of the dentists nor the standard errors of the estimates are much changed. The estimates of the specificities are, again, fairly similar but their precision has gone down. The standard errors are, on the whole, larger. Obviously the quality of the estimates gets better (that is, closer to those obtained from the full data and with smaller standard errors) as we go from Plan A to B to C. Using the naïve complete case approach does not lead us too far astray, except for the dramatic impact on the estimate of sensitivity for Dentist E. The estimate of the prevalence of caries will also be biased (too high) when we use the complete case approach.

5.11 Summary

Although we have looked briefly at patterns of agreement for binary ratings in this chapter, we have concentrated on the use of latent class models to attempt to explain and understand these patterns. Although most of the published work on the evaluation of diagnostic tests is based on having access to a definitive ('true' or 'gold standard') diagnosis, the analyses presented here are based on the assumption that most of the time we do not have access to the truth. If, however, we can get true values, then our inferences can become much stronger. We are then in an analogous situation to having access to a ruler when assessing the characteristics of subjective estimates of the lengths of pieces of string (Table 1.2). Occasionally there will be prior information on the characteristics of one or more of the diagnostic tests from earlier comparisons with a gold standard (see, for example, Gart & Buck, 1966). Again, this can be a great help. Even when we do have access to supposed gold standard assessments we can always relax the assumption that they are infallible in our models (again, a suitable analogy is the ruler and guesses of the lengths of pieces of string – see Section 4.7). Some authors (see, for example, Sullivan Pepe, 2003, page 203) are quite critical of the use of latent class analysis in the valuation of diagnostic tests. They state, quite correctly, that the latent class postulated within a latent class model may not actually correspond to a definition of illness or health in a clinical sense. It is simply a statistical device to explain the associations between a set of fallible tests. This is true and is also true of all of the other measurement error models presented in this book. Although it is, indeed, a scientific problem, it is also one of the great strengths of statistical modelling approach. It forces us to make our assumptions explicit and, ideally, it should stimulate us to design studies which will enable us to explore the sensitivity of our inferences to the sorts of assumptions we might be prepared to make. The fact that the differing assumptions built into the various analyses of the Gart and Buck syphilis test data (Section 5.8) lead to different conclusions should not lead us to decide that one approach is the correct one and that the other is misguided. The possibility of conflicting results should be the challenge for us to find an explanation.

Appendix 1: *EQS* programs (Chapter 3)

(a) EQS listing for 'orthogonal' regression

```
/TITLE
 CATSCAN DATA
/SPECIFICATION
 CASES=50;
 VARIABLES=2;
 METHOD=ML;
 MATRIX=CORRELATION;
 ANALYSIS=MOMENT;
/LABELS
 V1=PIXMEAN;
 V2=PLANMEAN;
/EQUATIONS
 V1=*V999+*F1+E1;
 V2=F1+E2;
 F1=*V999+D1;
/VARIANCES
 D1=2*;
 E1=0.002*;
 E2=0.049*;
/CONSTRAINTS
 0.041(E2,E2)-(E1,E1)=0;
/MATRIX
 1.000
 0.7029 1.000
/STANDARD DEVIATIONS
 0.5228 0.3969
/MEANS
 1.4079 1.7864
/END
```

(b) EQS listing for regression with known σ_δ^2

```
/TITLE
 CATSCAN DATA
/SPECIFICATION
 CASES=50;
 VARIABLES=2;
 METHOD=ML;
 MATRIX=CORRELATION;
 ANALYSIS=MOMENT;
/LABELS
 V1=PIXMEAN;
 V2=PLANMEAN;
/EQUATIONS
 V1=*V999+*F1+E1;
 V2=F1+E2;
 F1=*V999+D1;
/VARIANCES
 D1=2*;
 E1=0.002*;
 E2=0.025;
/MATRIX
 1.000
 0.7029 1.000
/STANDARD DEVIATIONS
 0.5228 0.3969
/MEANS
 1.4079 1.7864
/END
```

(c) EQS listing with known σ_δ^2 and $\beta=1$

```
/TITLE
 CATSCAN DATA
/SPECIFICATION
 CASES=50;
 VARIABLES=2;
 METHOD=ML;
 MATRIX=CORRELATION;
 ANALYSIS=MOMENT;
/LABELS
 V1=PIXMEAN;
 V2=PLANMEAN;
/EQUATIONS
 V1=*V999+F1+E1;
```

```
V2=F1+E2;
F1=*V999+D1;
/VARIANCES
D1=2*;
E1=0.002*;
E2=0.025;
/MATRIX
1.000
0.7029 1.000
/STANDARD DEVIATIONS
0.5228 0.3969
/MEANS
1.4079 1.7864
/END
```

(d) EQS listing for Grubbs' estimators

```
/TITLE
CATSCAN DATA
/SPECIFICATION
CASES=50;
VARIABLES=2;
METHOD=ML;
MATRIX=CORRELATION;
ANALYSIS=MOMENT;
/LABELS
V1=PIXMEAN;
V2=PLANMEAN;
/EQUATIONS
V1=*V999+F1+E1;
V2=F1+E2;
F1=*V999+D1;
/VARIANCES
D1=2*;
E1=0.002*;
E2=0.025*;
/MATRIX
1.000
0.7029 1.000
/STANDARD DEVIATIONS
0.5228 0.3969
/MEANS
1.4079 1.7864
/END
```

(e) EQS run to constrain error variances to be equal (β also assumed to be 1)

```
/TITLE
 CATSCAN DATA
/SPECIFICATION
 CASES=50;
 VARIABLES=2;
 METHOD=ML;
 MATRIX=CORRELATION;
 ANALYSIS=MOMENT;
/LABELS
 V1=PIXMEAN;
 V2=PLANMEAN;
/EQUATIONS
 V1=*V999+F1+E1;
 V2=F1+E2;
 F1=*V999+D1;
/VARIANCES
 D1=2*;
 E1=0.02*;
 E2=0.02*;
/constraints
 (E1,E1)=(E2,E2);
/MATRIX
 1.000
 0.7029 1.000
/STANDARD DEVIATIONS
 0.5228 0.3969
/MEANS
 1.4079 1.7864
/END
```

Appendix 2: Selected *EQS* programs (Chapter 4)

(a) Set up to fit measurement model for string data, with zero intercepts

```
/TITLE
 Measurement model for string data
 with zero intercepts
/SPECIFICATION
 CASES=15;
 VARIABLES=4;
 METHOD=ML;
 MATRIX=CORRELATION;
 ANALYSIS=MOMENT;
/LABELS
 V1=RULER;
 V2=GRAHAM;
 V3=BRIAN;
 V4=ANDREW;
 D1=LENGTH;
/EQUATIONS
 V2=F1+E2;
 V3=1*F1+E3;
 V4=1*F1+E4;
 F1=*V999+D1;
/VARIANCES
 D1=4*;
 E2 TO E4=0.1*;
/MATRIX
 1.000
 0.9802 1.000
```

```
 0.9811 0.9553 1.000
 0.9899 0.9807 0.9684 1.000
/MEANS
 4.2933 3.4333 3.4267 3.7667
/STANDARD DEVIATIONS
 1.9692 1.4563 1.6294 1.7987
/END
```

(b) Set up of a multitrait multimethod model for depression and anxiety

```
/TITLE
 Multitrait-multimethod analyses
 Depression & Anxiety data - Cole et al. (1997)
 MODIFIED SCALES
/SPECIFICATION
 CASES=280;
 VARIABLES=8;
 METHOD=ML;
 MATRIX=CORRELATION;
 ANALYSIS=COVARIANCE;
/LABELS
 V1=D-SELF; V2=D-PEER; V3=D-TEACH; V4=D-PARENT;
 V5=A-SELF; V6=A-PEER; V7=A-TEACH; V8=A-PARENT;
 F1=DEPRESSION;
 F2=ANXIETY;
 F3=SELF;
 F4=PEER;
 F5=TEACH;
 F6=PARENT;
/EQUATIONS
 V1=1.3*F1+6.5*F3+E1;
 V2=0.3*F1+0.3*F4+E2;
 V3=1.3*F1+5.0*F5+E3;
 V4=1.5*F1+4.1*F6+E4;
 V5=2.2*F2+6.5*F3+E5;
 V6=0.1*F2+0.3*F4+E6;
 V7=0.3*F2+5.0*F5+E7;
 V8=3.1*F2+4.1*F6+E8;
/VARIANCES
 F1 TO F2=1.0;
 F3 TO F6=1.0;
 E1 TO E8=3*;
/COVARIANCES
 F1,F2=0.81*;
```

```
/CONSTRAINTS
 (V1,F3)=(V5,F3);
 (V2,F4)=(V6,F4);
 (V3,F5)=(V7,F5);
 (V4,F6)=(V8,F6);
/MATRIX
 1.00
 0.21 1.00
 0.19 0.32 1.00
 0.21 0.32 0.27 1.00
 0.66 0.17 0.10 0.16 1.00
 0.18 0.72 0.26 0.22 0.10 1.00
 0.15 0.25 0.76 0.25 0.06 0.27 1.00
 0.24 0.33 0.22 0.72 0.22 0.25 0.20 1.00
/STANDARD DEVIATIONS
 7.21 0.43 6.02 4.56 9.44 0.39 5.51 6.24
/END
```

Appendix 3: Selected *gllamm* programs

(a) Measurement model for string data (including Rule)

```
generate one=1
eq f1: one
eq load: m1 m2 m3 m4
eq heter: m1 m2 m3 m4
gllamm y m2 m3 m4, i(string) eqs(load) s(heter) nocons adap
nip(15) geqs(f1) trace
```

Notes: the string's measured length is the variable **y** and the identification number (1 to 15) for the piece of string is the variable **string**. **m1** to **m4** are binary dummy variables (m1 = 1 for Rule, 0 otherwise; m2 = 1 for Graham, 0 otherwise; m3 = 1 for Brian, 0 otherwise and m4 = 1 for Andrew, 0 otherwise).

(b) Measurement model for testicle volumes (with error variances increasing with volume)

```
generate one=1
eq f1: one
eq load: m1 m2 m3 m4 m5
eq heter: m1 m2 m3 m4 m5 lmeancen
gllamm vol m2 m3 m4 m5, i(subject) eqs(load) s(heter) nocons
adap nip(15) geqs(f1) trace
```

Notes: the observed testicle volume is indicated by the variable **vol** and the identification number of the boy is the variable **subject**. **m1** to **m5** are binary dummy variables indicating the method of measurement. The variable **lmeancen** is the centred mean of the natural logarithms of the five volume measurements.

Appendix 4: Selected *Mplus* programs

(a) Single group model for CIS-R data

```
DATA:       FILE IS CIS-R.DAT;
VARIABLE:   NAMES ARE GHQ HADS CISR;
            USEVARIABLES GHQ HADS CISR;
ANALYSIS:   TYPE=MEANSTRUCTURE;
MODEL:

            F1 BY GHQ HADS CISR;
            [GHQ@0 HADS*0 CISR*0 F1];
            GHQ HADS CISR;
```

(b) Finite mixture model for CIS-R data (error variances equal for the two groups)

```
DATA:       FILE IS CIS-R.DAT;
VARIABLE:   NAMES ARE GHQ HADS CISR;
            CLASSES C(2);
            USEVARIABLES GHQ HADS CISR;
ANALYSIS:   TYPE=MIXTURE MISSING;
            ESTIMATOR=MLF;
MODEL:

            %OVERALL%
            F1 BY GHQ HADS CISR;
            [GHQ@0]
            [HADS*0]  (1)
            [CISR*0]  (2)
            [F1];

            %C#1%
```

```
[F1*-.1]
[HADS*0]  (1)
[CIS1*0]  (2);
F1*5;
```

(c) Finite mixture model for CIS-R data (unconstrained error variances)

```
DATA:       FILE IS CIS-R.DAT;
VARIABLE:   NAMES ARE GHQ HADS CISR;
            CLASSES C(2);
            USEVARIABLES GHQ HADS CISR;
ANALYSIS:   TYPE=MIXTURE MISSING;
            ESTIMATOR=MLF;
MODEL:
            %OVERALL%
            F1 BY GHQ HADS CISR;
            [GHQ@0]
            [HADS*0]  (1)
            [CISR*0]  (2)
            [F1];

            %C#1%
            [F1*-.1]
            [HADS*0]  (1)
            [CIS1*0]  (2);
            F1*5;
            GHQ*1 HADS*1 CISR*1;
```

Appendix 5: Selected *Mplus* runs to fit latent class models

(a) Latent class model for Holmquist biopsy data*

```
DATA:        FILE IS C:\Holmquist.dat
VARIABLE:    NAMES A B C D E F G;
             CLASSES C(2);
             CATEGORICAL A B C D E F G;
             USEVARIABLES A B C D E F G;
ANALYSIS:    TYPE=MIXTURE;
MODEL:
             %OVERALL%

             %C#1%
             [A$1*3 B$1*3 C$1*3 D$1*3 E$1*3 F$1*3 G$1*3];

             %C#2%
             [A$1*-2  B$1*-2  C$1*-2  D$1*-2  E$1*-2  F$1*-2
             G$1*-2];
OUTPUT:
             TECH10;
```

* Using a dataset with one record per biopsy slide (i.e. a row of seven binary 0/1 variables per case).

(b) Latent class model for Holmquist biopsy data, with constraints

```
DATA:        FILE IS C:\Holmquist.dat
VARIABLE:    NAMES A B C D E F G;
```

```
            CLASSES  C(2);
            CATEGORICAL  A  B  C  D  E  F  G;
            USEVARIABLES  A  B  C  D  E  F  G;
ANALYSIS:   TYPE=MIXTURE;
MODEL:

            %OVERALL%

            %C#1%
            [A$1*3  B$1*3]  (1);
            [C$1*3  D$1*3]  (2);
            [E$1*3]  (1);
            [F$1*3]  (2)  ;
            [G$1*3]  (1);

            %C#2%
            [A$1*-2  B$1*-2]  (3);
            [C$1*-2  D$1*-2]  (4);
            [E$1*-2]  (3);
            [F$1*-2]  (4)
            [G$1*-2]  (3);
OUTPUT:

            TECH10;
```

(c) Latent class model, with an instrumental variable, for Gart & Buck syphilis test data*

```
DATA:       FILE IS C:\Syphilis.dat;
VARIABLE:   NAMES AGEGP FTA VDRL A2 A3 A4 A5;
            CLASSES C(2);
            CATEGORICAL FTA VDRL;
            USEVARIABLES FTA VDRL A2 A3 A4 A5;
ANALYSIS:   TYPE=MIXTURE;
MODEL:

            %OVERALL%
            C#1 ON A2 A3 A4 A5;

            %C#1%
            [FTA$1@2.1972 VDRL$1*3];

            %C#2%
            [FTA$1@-2.9443 VDRL$1*-3];
```

* AGEGP is a variable coded from 1 to 5 (treated as a quantitative covariate in other analyses). A2 to A5 are binary dummy variables to distinguish the five age groups (when treating age group as a qualitative factor, as in the present run).

Appendix 6: Web-based resources and software packages

It would seem to be a pointless exercise to attempt to list and review all of the software packages that are presently available which might be useful in various forms of measurement error modelling. But I will try to indicate the range of possibilities. Like me, readers might find that they can find virtually all of the information they need from the internet. In using a search engine to find suitable material, I have found that the following three key phrases have been particularly useful: 'structural equation models', 'latent class analysis' and 'multilevel models'. Other key phrases such as 'reliability of measurements', 'measurement error models' and 'generalized mixed models' also turn up useful material. At the time of writing, the following three websites are an ideal place to start in the search for information on software and other practicalities.

Ed Rigdon's pages on structural equation modelling:

http://www.gsu.edu/~mkteer/index.html

John Uebersax's pages on latent class analysis:

http://ourworld.compuserve.com/homepages/jsuebersax/

The Centre for Multilevel Modelling:

http://multilevel.ioe.ac.uk/index.html

The better-known structural equation/covariance structure modelling packages (for the material in Chapters 3 and 4) include *Amos* (part of *SPSS*), *Proc Calis* (part of *SAS*), *EQS*, *LISREL*, *Mx*, and *MPlus*. Software for generalized mixed models (Proc *NLMIXED* in *SAS* and *gllamm* in *Stata*, for example) is also suitable here. There are also *sem* routines in *S-PLUS* and *R*. It is quite straightforward to find information on any of these products through the Web.

In the present text I have used *EQS* (Bentler, 1995), *Mplus* (Muthén & Muthén, 1998–2001) and *gllamm* (Rabe-Hesketh *et al.*, 2002) to fit measurement error models of various complexities to quantitative data. Further information on these three programs can be found on the following three websites, respectively:

http://www.mvsoft.com/
http://www.StatModel.com
http://www.gllamm.org

All routine data analyses and production of the figures in the present text have been carried out using *Stata* (StataCorp, 2003).

Packages for latent class analysis include *Latent GOLD*, *PanMark*, *LEM*, *Mplus*, and many more. *gllamm* can also fit latent class models. Again, it is easy to find further information on their respective websites. In the present text I have used *Latent GOLD* (Magidson & Vermunt, 2000), *MPlus* (see above) and *LATCLASS* (see Bartholomew & Knott, 1999). Further information on *Latent GOLD* and *LATCLASS* can be found at the following two sites, respectively:

http://www.statisticalinnovations.com
http://www.arnoldpublishers.com/support/lvmfa2.htm

Software for analysis of variance is too familiar to need any description, and most general purpose packages have variance components programs. Further information (including a series of reviews) on software packages for more complex multilevel (hierarchical) or mixed models can be found on the Centre for Multilevel Modelling web pages, above.

References and select bibliography

This bibliography includes all references cited in the text together with many others that are included for further reading. It is intended that the latter should complement the text and provide a fairly comprehensive entry to the literature on the evaluation of measurement errors. However, no attempt has been made to provide a complete list of references in the field. Several areas of interest that are not covered in great detail in the text can be followed from these references. The reader, for example, will be able to find several papers on formal sample size and power calculations in the list below, but this topic is only very briefly discussed in the text. Other areas include much of the work on the assessment and modelling of patterns of agreement for categorical measurements. The selection of references is inevitably a case of personal choice but I hope that this is sufficiently broad to satisfy most of the readers of the book.

Adcock, R.J. (1877). On the method of least squares. *The Analyst* **4**, 183–4.

Adcock, R.J. (1878). A problem in least squares. *The Analyst* **5**, 53–5.

Agresti, A. (1988). A model for agreement between ratings on an ordinal scale. *Biometrics* **44**, 539–48.

Agresti, A. (1989). An agreement model with kappa as a parameter. *Statistics & Probability Letters* **7**, 271–3.

Agresti, A. (1992). Modelling patterns of agreement and disagreement. *Statistical Methods in Medical Research* **1**, 201–18.

Agresti, A. (1999). Modelling ordered categorical data: recent advances and future challenges. *Statistics in Medicine* **18**, 2191–207.

Agresti, A. (2002). *Categorical Data Analysis* (2nd Edition). New York: Wiley.

Agresti, A. & Lang, J.B. (1993). Quasi-symetric latent class models, with application to rater agreement. *Biometrics* **49**, 131–9.

Agresti, A. & Yang, M.-C. (1987). An empirical investigation of some effects of sparseness in contingency tables. *Computational Statistics & Data Analysis* **5**, 9–21.

Aickin, M. (1990). Maximum likelihood estimation of agreement in the constant predictive probability model, and its relation to Cohen's kappa. *Biometrics* **46**, 293–302.

Aitkin, M., Anderson, D. & Hinde, J. (1981). Statistical modelling of data on teaching styles. *Journal of the Royal Statistical Society Series A* **144**, 419–61.

Alanen, E., Leskinen, E. & Kuusinen, J. (1998). Testing equality of reliability and stability with simple linear constraints in multi-wave, multi-variable models. *British Journal of Mathematical and Statistical Psychology* **51**, 327–41.

Albert, P., McShane, L.M., Shih, J.H. and The U.S. National Cancer Institute Bladder Tumor Marker Network. (2001). Latent class modeling approaches for assessing diagnostic error without a gold standard: with applications to p53 immunohistochemical assays in bladder tumors. *Biometrics* **57**, 610–19.

Allison, P.D. (1978). The reliability of variables measured as a number of events in an interval of time. In K.F. Schuessler (Ed.), *Sociological Methodology 1978*, 238–53. San Francisco: Jossey Bass.

Alswalmeh, Y.M. & Feldt, L.S. (1994). Testing the equality of two related intraclass reliability coefficients. *Applied Psychological Measurement* **18**, 183–90.

Altaye, M., Donner, A. & Eliasziw, M. (2001). A general goodness-of-fit approach for inference procedures concerning the kappa statistic. *Statistics in Medicine* **20**, 2479–88.

Altaye, M., Donner, A. & Klar, N. (2001). Inference procedures for assessing interobserver agreement among multiple raters. *Biometrics* **57**, 584–8.

Altman, D.G. & Bland, J.M. (1983). Measurement in medicine: the analysis of method comparison studies. *The Statistician* **32**, 307–17.

Alwin, D.F. & Jackson, D.J. (1980). Measurement models for response errors in surveys: issues and applications. In K.F. Schuessler (Ed.), *Sociological Methodology 1980*, 68–119. San Francisco: Jossey Bass.

Andersen, S. (1997). Re: 'Bayesian estimation of disease prevalence and the parameters of diagnostic tests in the absence of a gold standard.' *American Journal of Epidemiology* **145**, 290.

Anderson, D.A. & Aitkin, M. (1985). Variance components models with binary response: interview variability. *Journal of the Royal Statistical Society, Series B* **47**, 203–10.

Armitage, P., Berry, G. & Matthews, J.N.S. (2002). *Statistical Methods in Medical Research* (4th Edition). Oxford: Blackwell.

Arteaga, C., Jeyaratnam, S. & Graybill, F.A. (1982). Confidence intervals for proportions of total variance in the two-way components of variance model. *Communications in Statistics – Theory and Methods* **11**, 1643–58.

Baker, S.G. (1995). Evaluating multiple diagnostic tests with partial verification. *Biometrics* **51**, 330–7.

Baker, S., Freedman, L.S. & Parmar, M.K.B. (1991). Using replicate observations in observer agreement studies with binary assessments. *Biometrics* **47**, 1327–38.

Banerjee, M., Capozzoli, M., McSweeney, L. & Sinha, D. (1999). Beyond kappa: a review of interrater agreement measures. *The Canadian Journal of Statistics* **27**, 3–23.

Barnett, V.D. (1969). Simultaneous pairwise linear structural relationships. *Biometrics* **25**, 129–42.

Barnett, V.D. (1970). Fitting straight lines – the linear functional relationship with replicated observations. *Applied Statistics* **19**, 135–44.

Barnett, V.D. (1987). Straight consulting. In D.J. Hand & B.S. Everitt (Eds) *The Statistical Consultant in Action*, 26–41. Cambridge: Cambridge University Press.

Barnhart, H.X., Haber, M. & Song, J. (2002). Overall concordance correlation coefficient for evaluating agreement among multiple observers. *Biometrics* **58**, 1020–27.

Barnhart, H.X. & Williamson, J.M. (2001). Modelling concordance correlation via GEE to evaluate reproducibility. *Biometrics* **57**, 931–40.

Bartfay, E. & Donner, A. (2001). Statistical inferences for interobserver agreement studies with nominal outcome data. *The Statistician* **50**, 135–46.

Bartholomew, D.J. (1995). Spearman and the origin and development of factor analysis. *British Journal of Mathematical and Statistical Psychology* **48**, 211–20.

Bartholomew, D.J. (1996). *The Statistical Approach to Social Measurement*. San Diego, CA: Academic Press.

Bartholomew, D.J. & Knott, M. (1999). *Latent Variable Models and Factor Analysis* (2nd Edition). London: Arnold.

Bartholomew, D.J. & Leung, S.O. (2002). A goodness of fit test for sparse 2(p) contingency tables. *British Journal of Mathematical & Statistical Psychology* **55**, 1–15.

Bartko, J.J. (1966). The intraclass correlation coefficient as a measure of reliability. *Psychological Reports* **19**, 3–11.

Bartko, J.J. (1976). On various intraclass correlation reliability coefficients. *Psychological Bulletin* **83**, 762–5.

Bartko, J.J. (1994). General methodology II – Measures of agreement: a single procedure. *Statistics in Medicine* **13**, 737–45.

Bassein, L., Borghi, C., Costa, F.V., Strochi, E., Mussi, A. & Ambrosioni, E. (1985). Comparison of three devices for measuring blood pressure. *Statistics in Medicine* **4**, 361–8.

Basu, S., Basu, A. & Raychaudhuri, A. (1999). Measuring agreement between two raters for ordinal response: a model-based approach. *The Statistician* **48**, 339–48.

Basu, S., Raychaudhuri, A. & Basu, A. (1995). Improving precision through modelling: an illustration with hierarchical kappa. *Communications in Statistics – Simulation* **24**, 399–408.

Batista-Foguet, J.M., Coenders, G. & Ferragud, M.A. (2001), Using structural equation models to evaluate the magnitude of measurement error in blood pressure. *Statistics in Medicine* **20**, 2351–68.

BCA Commission on Nutritional Epidemiology (1993). Recommendations for the design and analysis of nutritional epidemiologic studies with measurement errors in the exposure variables. *European Journal of Clinical Nutrition* **47**, Suppl. 2, S53–7.

Becker, M.P. (1989). Using association models to analyse agreement data: two examples. *Statistics in Medicine* **8**, 1199–207.

Becker, M.P. & Agresti, A. (1992). Log-linear modelling of pairwise interobserver agreement on a categorical scale. *Statistics in Medicine* **11**, 101–14.

Becker, M.P. & Yang, I. (1998). Latent class marginal models for cross-classifications of counts. In A.E. Raftery (Ed.), *Sociological Methodology* **28**, 293–325. Boston & Oxford: Blackwell.

Begg, C.B. & Greene, R.A. (1983). Assessment of diagnostic tests when disease verification is subject to selection bias. *Biometrics* **39**, 207–15.

Bentler, P.M. (1995). *EQS Structural Equations Program Manual*. Encino, CA: Multivariate Software, Inc.

Bergan, J.R. (1980). Measuring observer agreement using the quasi-independence concept. *Journal of Educational Measurement* **17**, 59–69.

Bergan, J.R. (1980). A quasi-independence model for measuring observer agreement. *Journal of Educational Statistics* **5**, 363–76.

Berry, G., Smith, C.L., Macaskill, P. & Irwig, L. (2002). Analytic methods for comparing two dichotomous screening or diagnostic tests applied to two populations of differing disease prevalence when individuals neagative on both tests are unverified. *Statistics in Medicine* **21**, 853–62.

Besag, F.M.C., Hills, M., Wardale, F., Andrew, C.M. & Craggs, M.D. (1989). The validation of a new ambulatory spike and wave monitor. *Electroencephalography and Clinical Neurophysiology* **73**, 157–61.

Bhatia, A., Mangat, N.S. & Morrison, T. (1998). Estimation of measurement errors. *Proceedings of the International Pipeline Conference* 1998, Volume 1, 315–25. New York: The American Society of Mechanical Engineers.

Biemer, P.B. & Wiessan, C. (2002). Measurement error evaluation of self-reported drug use: a latent class analysis of the US National Household Survey on Drug Abuse. *Journal of the Royal Statistical Society, Series A* **165**, 97–119.

Bishop, Y.M.M., Fienberg, S.E. & Holland, P.W. (1975). *Discrete Multivariate Analysis: Theory and Practice*. Cambridge, Mass.: MIT Press.

Blackman, N.J.-M. & Koval, J.J. (2000). Interval estimation for Cohen's kappa as a measure of agreement. *Statistics in Medicine* **19**, 723–41.

Bland, J.M. & Altman, D.G. (1986). Statistical methods for assessing agreement between two methods of clinical measurement. *Lancet* **i**, 307–10.

Bland, J.M. & Altman, D.G. (1995). Comparing two methods of clinical measurement: a personal history. *International Journal of Epidemiology* **24** (Suppl. 1), S7–14.

Bland, J.M. & Altman, D.G. (1996). Measurement error. *British Medical Journal* **313**, 744.

Bland, J.M. & Altman, D.G. (1999). Measuring agreement in method comparison studies. *Statistical Methods in Medical Research* **8**, 135–60.

Bloch, D.A. & Kraemer, H.C. (1989). 2×2 kappa coefficients: measures of agreement or association. *Biometrics* **45**, 269–87.

Bodian, C.A. (1994). Intraclass correlation for two-by-two tables under three sampling designs. *Biometrics* **50**, 183–93.

Bogle, W. & Hsu, Y.S. (2002). Sample size determination in comparing two population variances with paired data: application to bilirubin tests. *Biometrical Journal* **44**, 594–602.

Bolfarine, H., Cabral, C.R.B. & Paula, G.A. (2002). Distance tests under nonregular conditions: applications to the comparative calibration model. *Journal of Statistical Computation and Simulation* **72**, 125–40.

Bolfarine, H. & Galea-Rojas, M. (1995). Comments on 'Functional Comparative Calibration Using an EM Algorithm' by D. Kimura. *Biometrics* **51**, 1579–80.

Bolfarine, H. & Galea-Rojas, M. (1995). Structural comparative calibration using the EM algorithm. *Journal of Applied Statistics* **22**, 277–90.

Bolfarine, H. & Galea-Rojas, M. (1995). Maximum likelihood estimation of simultaneous pairwise linear structural relationships. *Biometrical Journal* **37**, 673–89.

Bolfarine, H. & Galea-Rojas, M. (1996). On structural comparative calibration under a t-model. *Computational Statistics* **11**, 63–85.

Bolfarine, H. & Vilca-Labra, F. (2000). Letter to the Editor on 'Comparative Calibration Without a Gold Standard' by Lu *et al. Statistics in Medicine* **19**, 1253–4.

Bollen, K.A. (1989). *Structural Equation Models with Latent Variables*. New York: Wiley.

Bollen, K.A. & Lennox, R. (1991). Conventional wisdom on measurement: a structural equation perspective. *Psychological Bulletin* **110**, 305–14.

Bonett, D.G. (2002). Sample size requirements for estimating intraclass correlations with desired precision. *Statistics in Medicine* **21**, 1331–5.

Bost, J.E. (1995). The effects of correlated errors on generalizability and dependability coefficients. *Applied Psychological Measurement* **19**, 191–203.

Bradley, E.L. & Blackwood, L.G. (1989). Comparing paired data: a simultaneous test for means and variances. *The American Statistician* **43**, 234–5.

Brennan, R.L. (2001). *Generalizability Theory*. New York: Springer.

Brennan, R.L., Harris, D.J. & Hanson, B.A. (1987). *The bootstrap and other procedures for examining the variability of estimated variance components in testing contexts* (American College Testing Research report No. 87-7). Iowa city, IA: ACT, Inc.

Browne, M.W. & Arminger, G. (1995). Specification and estimation of mean- and covariance-structure models. In G. Arminger, C.C. Clogg & M.E. Sobel (Eds), *Handbook of Statistical Modelling for the Social and Behavioural Sciences* 311–59 New York: Plenum Press.

Browne, M.W. & Shapiro, A. (1988). Robustness of normal theory methods in the analysis of latent variable models. *British Journal of Mathematical and Statistical Psychology* **41**, 193–208.

Burch, B.D. & Harris, I.R. (2001). Closed-form approximations to the REML estimator of a variance ratio (or heritability) in a mixed linear model. *Biometrics* **57**, 1148–56.

Burdick, R.K. & Graybill, F.A. (1992). *Confidence Limits on Variance Components*. New York: Dekker.

Bush, N. & Anderson, R.L. (1963). A comparison of three different procedures for estimating variance components. *Technometrics* **5**, 421–40.

Campbell, D.T. & Fiske, D.W. (1959). Convergent and discriminant validation by the multitrait-multimethod matrix. *Psychological Bulletin* **56**, 81–105.

Cantor, A. (1996). Sample-size calculations for Cohen's kappa. *Psychological Methods* **1**, 150–3.

Cappelleri, J.C. & Ting, N. (2003). A modified large-sample approach to approximate interval estimation for a particular intraclass correlation coefficient. *Statistics in Medicine* **22**, 1861–77.

Carey, B.L. (1993). Measurement assurance: role of statistics and support from international statistical standards. *International Statistical Review* **61**, 27–40.

Carmines, E.G. & Zeller, R.A. (1979). *Reliability and Validity Assessment*. Thousand Oaks: Sage.

Carpenter, J. & Bithell, J. (2000). Bootstrap confidence intervals: when, which, what? A practical guide for medical statisticians. *Statistics in Medicine* **19**, 1141–64.

Carroll, R.J. (2003). Variances are not always nuisance parameters. *Biometrics* **59**, 211–20.

Carroll, R.J., Freedman, L. & Pee, D. (1997). Design aspects of calibration studies in nutrition, with analysis of missing data in linear measurement error models. *Biometrics* **53**, 1440–57.

Carroll, R.J., Roeder, K. & Wasserman, L. (1999). Flexible parametric measurement error models. *Biometrics* **55**, 44–54.

Carroll, R.J. & Ruppert, D. (1988). *Transformation and Weighting in Regression*. New York: Chapman & Hall CRC Press.

Carroll, R.J. & Ruppert, D. (1996). The use and misuse of orthogonal regression in linear errors-in-variables models. *The American Statistician* **50**, 1–6.

Carroll, R.J., Ruppert, D. & Stefanski, L.A. (1995). *Measurement Error in Non-Linear Models*. London: Chapman & Hall.

Carter, R.L. (1981). Restricted maximum likelihood estimation of bias and reliability in the comparison of several measuring methods. *Biometrics* **37**, 733–41.

Caulcott, R. & Boddy, R. (1983). *Statistics for Analytical Chemists*. London: Chapman & Hall.

Cheng, C.-L. & Van Ness, J.W. (1997). Robust calibration. *Technometrics* **39**, 401–11.

Cheng, C.-L. & Van Ness, J.W. (1999). *Statistical Regression with Measurement Error*. London: Arnold.

Chinchilli, V.M., Martel, J.K., Kumanyika, S. & Lloyd, T. (1996). A weighted concordance correlation coefficient for repeated measurement designs. *Biometrics* **52**, 341–53.

Chinn, S. (1990). The assessment of methods of measurement. *Statistics in Medicine* **9**, 351–62.

Chipkevitch, E., Nishimuro, R.T., Tu, D.G.S. & Galea-Rojas, M. (1996). Clinical measurement of testicular volume in adolescents: comparison of the reliability of 5 methods. *The Journal of Urology* **156**, 2050–3.

Chock, C., Irwig, L., Berry, G. & Glasziou, P. (1997). Comparing dichotomous screening tests when individuals negative on both tests are not verified. *Journal of Clinical Epidemiology* **50**, 1211–7.

Christensen, R. & Blackwood, L.G. (1993). Tests of precision and accuracy of multiple measuring devices. *Technometrics* **35**, 411–20.

Clare, A.W. & Cairns, V.E. (1978). Design, development and the use of a standardized interview to assess social maladjustment and dysfunction in community studies. *Psychological Medicine* **8**, 589–604.

Clayton, D. (1985). Using test-retest reliability data to improve estimates of relative risk: an application of latent class analysis. *Statistics in Medicine* **4**, 445–55.

Clogg, C.C. & Goodman, L.A. (1985). Simultaneous latent structure analysis in several groups. In Tuma, N.B. (Ed.), *Sociological Methodology 1985*, 81–110. San Francisco: Jossey-Bass.

Cochran, W.G. (1943). The comparison of different scales of measurement for experimental results. *Annals of Mathematical Statistics* **14**, 205–16.

Cochran, W.G. (1968). Errors of measurement in statistics. *Technometrics* **10**, 637–66.

Cohen, J. (1960). A coefficient of agreement for nominal scales. *Educational and Psychological Measurement* **20**, 37–46.

Cohen, J. (1968). Weighted kappa: nominal scale agreement with provision for scales disagreement of partial credit. *Psychological Bulletin* **70**, 213–20.

Cole, D.A., Truglio, R. & Peeke, L. (1997). Relation between symptoms of anxiety and depression in children: a multitrait-multimethod-multigroup assessment. *Journal of Consulting and Clinical Psychology* **65**, 110–19.

Cole, D.A., Peeke, L.G., Martin, J.M., Truglio, R. & Seroczynski, A.D. (1998). A longitudinal look at the relation between depression and anxiety in children and adolescents. *Journal of Consulting and Clinical Psychology* **66**, 451–60.

Collins, L.M., Fidler, P.L., Wugalter, S.E. & Long, J.D. (1993). Goodness-of-fit for testing latent class models. *Multivariate Behavioral Research* **28**, 375–89.

Commenges, D. & Jacqmin, H. (1994). The intraclass correlation coefficient: distribution-free definition and test. *Biometrics* **50**, 517–26.

Conger, A.J. (1980). Integration and generalization of kappas for multiple raters. *Psychological Bulletin* **88**, 322–8.

Cook, R.J. & Farewell, V.T. (1995). Conditional inference for subject-specific and marginal agreement: two families of agreement measures. *The Canadian Journal of Statistics* **23**, 333–44.

Creasy, M.A. (1956). Confidence limits for the gradient in the linear functional relationship. *Journal of the Royal Statistical Society, Series B* **18**, 65–9.

Cronbach, L.J., Rajaratnam, N. & Gleser, G.C. (1963). Theory of generalizability: a liberation of reliability theory. *British Journal of Statistical Psychology* **16**, 137–63.

Cronbach, L.J., Gleser, G.L., Nanda, H. & Rajaratnam, N. (1972). *The Dependability of Behavioral Meaurements*. New York: Wiley.

Crowder, M. (1992). Interlaboratory comparisons: round robins with random effects. *Applied Statistics* **41**, 409–25.

Darroch, J.N. & McLoud, P.I. (1986). Category distinguishability and observer agreement. *Australian Journal of Statistics* **28**, 371–88.

Davidian, M., Carroll, R.J. & Smith, W. (1988). Variance functions and the minimum detectable concentration in assays. *Biometrika* **75**, 549–56.

Davies, M. & Fleiss, J.L. (1982). Measuring agreement for multinomial data. *Biometrics* **38**, 1047–51.

Davison, A.C. & Hinkley, D.V. (1997). *Bootstrap Methods and Their Application*. Cambridge: Cambridge University Press.

Dawid, A.P. & Skene, A.M. (1979). Maximum likelihood estimation of observer error-rates using the EM algorithm. *Applied Statistics* **28**, 20–8.

Dayton, C.M. & Macready, G.B. (1983). Latent structure analysis of repeated classifications with dichotomous data. *British Journal of Mathematical and Statistical Psychology* **36**, 189–201.

Deming, W.E. (1943). *Statistical Adjustment of Data*. New York: Wiley.

Dempster, A.P., Laird, N.M. & Rubin, D.B. (1977). Maximum likelihood from incomplete data via the *EM* algorithm. *Journal of the Royal Statistical Society, Series B* **39**, 1–38.

Dendukuri, N. & Joseph, L. (2001). Bayesian approaches to modelling the conditional dependence between multiple diagnostic tests. *Biometrics* **57**, 158–67.

Donner, A. (1986). A review of inference procedures for the intraclass correlation coefficient in the one-way random effects model. *International Statistical Review* **54**, 67–82.

Donner, A. (1998). Sample size requirements for the comparison of two or more coefficients of inter-observer agreement. *Statistics in Medicine* **17**, 1157–68.

Donner, A. (1999). Sample size requirements for interval estimation of the intraclass kappa statistic. *Communications in Statistics – Simulation* **28**, 415–29.

Donner, A. & Bull, S. (1983). Inferences concerning a common intraclass correlation coefficient. *Biometrics* **39**, 771–5.

Donner, A. & Eliasziw, M. (1994). Statistical implications of the choice between a dichotomous and a continuous trait in studies of interobserver reliability. *Biometrics* **50**, 550–5.

Donner, A. & Eliasziw, M. (1997). A hierarchical approach to inferences concerning interobserver agreement for multinomial data. *Statistics in Medicine* **16**, 1097–106.

Donner, A., Eliasziw, M. & Klar, N. (1996). Testing the homogeneity of kappa statistics. *Biometrics* **52**, 176–83.

Donner, A. & Klar, N. (1996). The statistical analysis of kappa statistics in multiple samples. *Journal of Clinical Epidemiology* **49**, 1053–8.

Donner, A., Shroukri, M.M., Klar, N. & Bartfay, E. (2000). Testing the equality of two dependent kappa statistics. *Statistics in Medicine* **19**, 373–87.

Donner, A. & Zou, G. (2002). Testing the equality of dependent intraclass correlation coefficients. *The Statistician* **51**, 367–79.

Donner, A. & Zou, G. (2002). Interval estimation for a difference between intraclass kappa statistics. *Biometrics* **58**, 209–15.

Dunn, G. (1989). *Design and Analysis of Reliability Studies*. London: Arnold.

Dunn, G. (1992). Design and analysis of reliability studies. *Statistical Methods in Medical Research* **1**, 123–57.

Dunn, G. (1998). Letter to the Editor on 'Comparative Calibration Without a Gold Standard' by Lu *et al*. *Statistics in Medicine* **17**, 1293–8.

Dunn, G. (2000). *Statistics in Psychiatry*. Arnold: London.

Dunn, G., Everitt, B. & Pickles, A. (1993). *Modelling Covariances and Latent Variables using EQS*. London: Chapman & Hall.

Dunn, G. & Roberts, C. (1999). Modelling method comparison data. *Statistical Methods in Medical Research* **8**, 161–79.

Dybkaer, R. (1997). Vocabulary for use in measurement procedures and description of reference materials in laboratory medicine. *European Journal of Clinical Chemistry and Clinical Biochemistry* **35**, 141–73.

Edlund, S.D. (1996). Bias in slope estimates for the linear errors in variables modelled by the variance ratio method. *Biometrics* **52**, 243–8.

Efron, B. (1979). Bootstrap methods: another look at the jackknife. *Annals of Statistics* **7**, 1–26.

Efron, B. & Tibshirani, R.J. (1993). *An Introduction to the Bootstrap*. London: Chapman & Hall.

Eksborg, S. (1981). Evaluation of method comparison data. *Clinical Chemistry* **27**, 1311–2.

Eliasziw, M. & Donner, A. (1987). A cost-function approach to the design of reliability studies. *Statistics in Medicine* **6**, 647–55.

Espeland, M.A. & Handelman, S.L. (1989). Using latent class models to characterize and assess relative error in discrete measurements. *Biometrics* **45**, 587–99.

Espeland, M.A., Platt, O.S. & Gallagher, D. (1989). Joint estimation of incidence and diagnostic error rates from irregular longitudinal data. *Journal of the American Statistical Association* **84**, 972–9.

Everitt, B.S. (1968). Moments of the statistics kappa and weighted kappa. *The British Journal of Mathematical and Statistical Psychology* **21**, 97–103.

Everitt, B.S. (1977). Some properties of statistics used for measuring observer agreement in the recording of signs. *The British Journal of Mathematical and Statistical Psychology* **30**, 227–33.

Everitt, B.S. & Hand, D.J. (1981). *Finite Mixture Distributions*. London: Chapman & Hall.

Fairfield Smith, H. (1950). Estimating precision of measuring instruments. *Journal of the American Statistical Association* **45**, 447–51.

Faraone, S.V. & Tsuang, M.T. (1994). Measuring diagnostic accuracy in the absense of a 'Gold Standard'. *American Journal of Psychiatry* **151**, 650–7.

Farewell, V.T. & Sprott, D.A. (1999). Conditional inference for predictive agreement. *Statistics in Medicine* **18**, 1435–49.

Feldstein, M.L. & Davis, H.T. (1984). Poisson models for assessing rater agreement in discrete response studies. *British Journal of Mathematical and Statistical Psychology* **37**, 49–61.

Feldt, L.S. (1965). The approximate sampling distribution of Kuder-Richardson reliability coefficient twenty. *Psychometrika* **30**, 357–70.

Feldt, L.S. & Brennan, R.L. (1993). Reliability. In R.L. Linn (Ed.) *Educational Measurement* (3rd Edition), 105–46. New York: American Council on Education and Macmillan.

Finney, D.J. (1996). A note on the history of regression. *Journal of Applied Statistics* **23**, 555–7.

Fisher, R.A. & Mackenzie, W.A. (1923). Studies in crop variation. II. The manural response of different potato varieties. *Journal of Agricultural Science* **13**, 311.

Fleiss, J.L. (1965). Estimating the accuracy of dichotomous judgements. *Psychometrika* **30**, 469–79.

Fleiss, J.L. (1970). Estimating the reliability of interview data. *Psychometrika* **35**, 143–62.

Fleiss, J.L. (1971). Measuring nominal scale agreement among many raters. *Psychological Bulletin* **76**, 378–82.

Fleiss, J.L. (1975). Measuring agreement between two judges on the presence or absence of a trait. *Biometrics* **31**, 651–9.

Fleiss, J.L. (1981a). Balanced incomplete blocks designs for inter-rater reliability studies. *Applied Psychological Measurement* **5**, 105–12.

Fleiss, J.L. (1981b). *Statistical Methods for Rates and Proportions* (2nd Edition). New York: Wiley & Sons.

Fleiss, J.L. (1986). *The Design and Analysis of Clinical Experiments*. New York: Wiley & Sons.

Fleiss, J.L. & Cohen, J. (1973). The equivalence of weighted kappa and the intraclass correlation coefficient as measures of reliability. *Educational and Psychological Measurement* **33**, 613–9.

Fleiss, J.L., Cohen, J. & Everitt, B.S. (1969). Large sample standard errors of kappa and weighted kappa. *Psychological Bulletin* **72**, 323–7.

Fleiss, J.L. & Cuzick, J. (1979). The reliability of dichotomous judgements: unequal numbers of judgements per subject. *Applied Psychological Measurement* **3**, 537–42.

Fleiss, J.L. & Shrout, P.E. (1978). Approximate interval estimation for a certain intraclass correlation coefficient. *Psychometrika* **43**, 259–62.

Formann, A.K. (2000). Rater agreement and the generalized Rudas-Clogg-Lindsay index of fit. *Statistics in Medicine* **19**, 1881–8.

Formann, A.K. (2003). Latent class model diagnosis from a frequentist point of view. *Biometrics* **59**, 189–96.

Formann, A.K. & Kohlmann, T. (1996). Latent class analysis in medical research. *Statistical Methods in Medical Research* **5**, 179–211.

Freedman, L.S., Carroll, R.J. & Wax, Y. (1991). Estimating the relation between dietary intake obtained from a food frequency questionnaire and true average intake. *American Journal of Epidemiology* **134**, 310–20.

Freedman, L.S., Parmar, M.K.B. & Baker, S.G. (1993). The design of observer agreement studies with binary assessments. *Statistics in Medicine* **12**, 165–79.

Fujisawa, H. & Izumi, S. (2000). Inference about misclassification probabilities from repeated binary responses. *Biometrics* **56**, 706–11.

Fuller, W.A. (1987). *Measurement Error Models*. New York: John Wiley & Sons.

Galea-Rojas, M., Bolfarine, H. & de Castro, M. (2002). Local influence in comparative calibration models. *Biometrical Journal* **44**, 59–81.

Garcia-Alfaro, K.H. & Bolfarine, H. (2001). Comparative calibration with subgroups. *Communications in Statistics – Theory and Methods* **30**, 2057–78.

Garrett, E.S., Eaton, W.W. & Zeger, S. (2002). Methods for evaluating the performance of diagnostic tests in the absence of a gold standard: a latent class model approach. *Statistics in Medicine* **21**, 1289–307.

Garrett, E.S. & Zeger, S.L. (2000). Latent class model diagnostics. *Biometrics* **56**, 1055–67.

Garrett, E.S. & Zeger, S.L. (2003). Rejoinder to Formann (2003). *Biometrics* **59**, 197–8.

Gart, J.J. & Buck, A.A. (1966). Comparison of a screening test and a reference test in epidemiological studies. II. A probabilistic model for the comparison of diagnostic tests. *American Journal of Epidemiology* **83**, 593–602.

Gaylor, D.W. (1960). *The construction and evaluation of some designs for the estimation of parameters in random models*. Unpublished PhD thesis. North Carolina State College, Raleigh, Institute of Statistics (Mimeo Series No. 256).

Georgiadis, M.P., Johnson, W., Gardner, I.A. & Singh, R. (2003). Correlation-adjusted estimation of sensitivity and specificity of two diagnostic tests. *Applied Statistics* **52**, 63–76.

Gibbons, R.D. & Bhaumik, D.K. (2001). Weighted random-effects regression models with application to interlaboratory calibration. *Technometrics* **43**, 192–8.

Gimenez, P. & Bolfarine, H. (2000). Comparing consistent estimators in comparative calibration models. *Journal of Statistical Planning and Inference* **86**, 143–55.

Gimenez, P., Bolfarine, H. & Colosimo, E.A. (2000). Hypothesis testing for error-in variables models. *Annals of the Institute of Statistical Mathematics* **52**, 698–711.

Giraudeau, B. & Mary, J.Y. (2001). Planning a reproducibility study: how many subjects and how many replicates per subject for an expected width of the 95 per cent confidence interval of the intraclass correlation coefficient. *Statistics in Medicine* **20**, 3205–14.

Giraudeau, B., Ravaud, P. & Chastang, C. (2000). Comment on Quan and Shih's 'Assessing reproducibility by a within-subject coefficient of variation with random effects models.' *Biometrics* **56**, 301–3.

Gleser, G.C., Cronbach, L.J. & Rajaratnam, N. (1965). Generalizability of scores influenced by multiple sources of variance. *Psychometrika* **30**, 395–418.

Gleser, G.C., Green, B.I. & Winget, C.N. (1978). Quantifying interview data on psychic impairment of disaster survivors. *Journal of Nervous and Mental Disease* **166**, 209–16.

Gleser, L.J. (1987). Confidence intervals for the slope in a linear errors-in-variables model. In A.K. Gupta (Ed.) *Advances in Multivariate Statistical Analysis*, 85–109. Dordrecht: D. Reidel.

Gleser, L.J. (1991). Measurement error models. *Chemometrics and Intelligent Laboratory Systems* **10**, 45–57.

Gleser, L.J. (1998). Assessing uncertainty in measurement. *Statistical Science* **13**, 277–90.

Gleser, L.J. & Hwang, J.T. (1987). The non-existence of $100(1 - \alpha)\%$ confidence sets of finite expected diameter in errors-in-variables and related models. *Annals of Statistics* **15**, 1351–62.

Goetghebeur, E., Liinev, J., Boelaert, M. & Van der Stuyft, P. (2000). Diagnostic test analyses in search of their gold standard: latent class analysis with random effects. *Statistical Methods in Medical Research* **9**, 231–48.

Goldberg, D.P. (1972). *The Detection of Psychiatric Illness by Questionnaire*. London: Oxford University Press.

Goldsmith, C.H. & Gaylor, D.W. (1970). Three stage nested designs for estimating variance components. *Technometrics* **12**, 487–98.

Goldstein, H. (1995). *Multilevel Statistical Models*. London: Arnold.

Gonin, R., Lipsitz, S.R., Fitzmaurice, G.M. & Molenberghs, G. (2000). Regression modelling of weighted κ by using generalized estimating equations. *Applied Statistics* **49**, 1–18.

Graham, P. & Jackson, R. (1993). The analysis of ordinal agreement data: beyond weighted kappa. *Journal of Clinical Epidemiology* **46**, 1055–62.

Grizzle, J.E., Starmer, C.F. & Koch, G.G. (1969). Analysis of categorical data by linear models. *Biometrics* **25**, 489–503.

Grubbs, F.E. (1948). On estimating precision of measuring instruments and product variability. *Journal of the American Statistical Association* **43**, 243–64.

Grubbs, F.E. (1973). Errors of measurement, precision, accuracy and statistical comparison of measuring instruments. *Technometrics* **15**, 53–66.

Guggenmoos-Holzmann, I. (1993). How reliable are chance-corrected measures of agreement? *Statistics in Medicine* **12**, 2191–205.

Guggenmoos-Holzmann, I. (1996). The meaning of kappa: probabilistic concepts of reliability and validity revisited. *Journal of Clinical Epidemiology* **49**, 775–82.

Guggenmoos-Holzmann, I. & van Houwelingen, H.C. (2000). The (in)validity of sensitivity and specificity. *Statistics in Medicine* **19**, 1783–92.

Guggenmoos-Holzmann, I. & Vonk, R. (1998). Kappa-like indices of observer agreement viewed from a latent class perspective. *Statistics in Medicine* **17**, 797–812.

Gustafson, P. (2003). *Measurement Error and Misclassification in Statistics and Epidemiology*. London: Chapman & Hall/CRC.

Haberman, S.J. (1979). *Analysis of Qualitative Data, Volume 2, New Developments*. New York: Academic Press.

Hadgu, A. & Qu, Y. (1998). A biomedical application of latent class models with random effects. *Applied Statistics* **47**, 603–16.

Hahn, G.J. & Nelson, W. (1970). A problem in the statistical comparison of measuring devices. *Technometrics* **12**, 95–102.

Hale, C.A. & Fleiss, J.L. (1993). Interval estimation under two study designs for kappa with binary classifications. *Biometrics* **49**, 523–34.

Hanamura, R.C. (1975). Estimating imprecisions of measuring instruments. *Technometrics* **17**, 299–302.

Hand, D.J. & Taylor, C.C. (1987). *Multivariate Analysis of Variance and Repeated Measures*. London: Chapman & Hall.

Hagenaars, J.A. & McCutcheon, A.L. (2002). *Applied Latent Class Analysis*. Cambridge: Cambridge University Press.

Hartley, H.O., Rao, J.N.K. & LaMotte, L.R. (1978). A simple synthesis-based method of estimating variance components. *Biometrics* **34**, 233–43.

Hartmann, C., Smeyers-Verbeke, J., Penninckx, W. & Massart, D.L. (1997). Detection of bias in method comparison by regression analysis. *Analytica Chimica Acta* **338**, 19–40.

Harville, D.A. (1977). Maximum likelihood approaches to variance component estimation and to related problems. *Journal of the American Statistical Association* **72**, 320–40.

Hawkins, D.M. (2002). Diagnostics for conformity of paired quantitative measurements. *Statistics in Medicine* **21**, 1913–35.

Hawkins, D.M., Garrett, J.A. & Stephenson, B. (2001). Some issues in resolution of diagnostic tests using an imperfect gold standard. *Statistics in Medicine* **20**, 1987–2001.

Healy, M.R.J. (1989). Measuring measurement errors. *Statistics in Medicine* **8**, 893–906.

Heise, D.R. (1969). Separating reliability and stability in test-retest correlation. *American Sociological Review* **34**, 93–101.

Heise, D.R. & Bohrnstedt, G.W. (1970). Validity, invalidity and reliability. In F. Borgatta & G.W. Bohrnstedt (Eds.), *Sociological Methodology 1970*, 104–29. San Francisco: Jossey Bass.

Henderson, C.R. (1953). Estimating variance and covariance components. *Biometrics* **9**, 226–52.

Hewitt, J.K., Silberg, J.L., Neale, M.C., Eaves, L.J. & Erickson, M. (1992). The analysis of parental ratings of children's behavior using LISREL. *Behavior Genetics* **22**, 293–317.

Heywood, H.B. (1931). On finite sequences of real numbers. *Proceedings of the Royal Society A* **134**, 486–510.

Holmquist, N.D., McMahon, C.A. & Williams, O.D. (1967). Variability in classification of carcinoma in situ of the uterine cervix. *Archives of Pathology* **84**, 334–45.

Hood, K., Nix, B.A.J. & Iles, T.C. (1999). Asymptotic information and variance-covariance matrices for the linear structural model. *The Statistician* **48**, 477–93.

Huber, P. (1967). The behavior of maximum likelihood estimates under non-standard conditions. In *Proceedings of the 5th Berkeley Symposium on Mathematical Statistics and Probability*, Vol. 1, 221–33. Berkeley: University of California Press.

Hubert, L. (1977a). Nominal scale agreement as a generalized correlation. *British Journal of Mathematical and Statistical Psychology* **30**, 98–103.

Hubert, L. (1977b). Kappa revisited. *Psychological Bulletin* **84**, 289–97.

Hubert, L. & Golledge, R.G. (1983). Rater agreement for complex assessments. *British Journal of Mathematical and Statistical Psychology* **36**, 207–16.

Hui, S.L. & Walter, S.D. (1980). Estimating the error rates of diagnostic tests. *Biometrics* **36**, 167–71.

Hui, S.L. & Zhou, X.H. (1998). Evaluation of diagnostic tests without gold standards. *Statistical Methods in Medical Research* **7**, 354–70.

International Standards Institute (1994a). *International Standard ISO 5725-1*. Accuracy (trueness and precision) of measurement methods and results. Part 1: General principles and definitions.

International Standards Institute (1994b). *International Standard ISO 5725-2*. Accuracy (trueness and precision) of measurement methods and results. Part 2: Basic method for the determination of the repeatability and reproducibility of a standard measurement method.

International Standards Institute (1994c). *International Standard ISO 5725-3*. Accuracy (trueness and precision) of measurement methods and results. Part 3: Intermediate measures of the precision of a standard measurement method.

International Standards Institute (1994d). *International Standard ISO 5725-4*. Accuracy (trueness and precision) of measurement methods and results. Part 4: Basic methods for determination of the trueness of a standard measurement method.

International Standards Institute (1994e). *International Standard ISO 5725-6*. Accuracy (trueness and precision) of measurement methods and results. Part 6: Use in practice of accuracy values.

International Standards Institute (1998). *International Standard ISO 5725-5*. Accuracy (trueness and precision) of measurement methods and results. Part 5: Alternative methods for the determination of the precision of a standard measurement method.

Jaech, J.L. (1979). Estimating within-laboratory variability from interlaboratory test data. *Journal of Quality Technology* **11**, 185–91.

Jaech, J.L. (1985). *Statistical Analysis of Measurement Errors*. New York: John Wiley & Sons.

Jagodzinski, K.G. & Kühnel, S.M. (1987). Estimation of reliability and stability in single-indicator multiple-wave panel models. *Sociological Methods & Research* **15**, 219–58.

Jagodzinski, K.G., Kühnel, S.M. & Schmidt, P. (1987). Is there a 'Socratic Effect' in nonexperimental panel studies? *Sociological Methods & Research* **15**, 259–302.

James, I.R. (1983). Analysis of nonagreements among multiple raters. *Biometrics* **39**, 651–7.

Jannarone, R.J., Macera, C.A. & Garrison, C.Z. (1987). Evaluating interrater agreement through 'Case-Control' sampling. *Biometrics* **43**, 433–7.

Jansen, A.A.M. (1980). Comparative calibration and congeneric measurements. *Biometrics* **36**, 729–34.

Janson, H. & Olsson, U. (2001). A measure of agreement for interval and nominal multivariate observations. *Educational and Psychological Measurement* **61**, 277–89.

Johnson, W.O. & Gastwith, J.L. (1991). Bayesian inference for medical screening tests: approximations useful for the analysis of acquired immunodeficiency syndrome. *Journal of the Royal Statistical Society, Series B* **53**, 427–39.

Johnson, W.O., Gastwith, J.L. & Pearson, L.M. (2001). Screening without a 'gold standard': the Hui-Walter paradigm revisited. *American Journal of Epidemiology* **153**, 921–4.

Jöreskog, K.G. (1971a). Statistical analysis of sets of congeneric tests. *Psychometrika* **36**, 109–35.

Jöreskog, K.G. (1971b). Simultaneous factor anlysis in several populations. *Psychometrika* **36**, 409–27.

Jöreskog, K.G. & Sörbom, D. (1979). *Advances in Factor Analysis and Structural Equations Models*. Cambridge, MA: Abt Books.

Jørgensen, B. (1985). Estimation of inter-observer variation for ordinal rating scales. In R. Gilchrist, B. Francis & J. Whittaker (Eds), *Generalized Linear Models*, 93–104. Berlin: Springer-Verlag.

Joseph, L., Gyorkos, T.W. & Coupal, L. (1995). Bayesian estimation of disease prevalence and the parameters of diagnostic tests in the absence of a gold standard. *American Journal of Epidemiology* **141**, 263–72.

Kaaks, R., Riboli, E., Esteve, J., Van Keppel, A.L. & Van Staveren, W.A. (1994). Estimating the accuracy of dietary questionnaire assessments: validation in terms of structural equation models. *Statistics in Medicine* **13**, 127–42.

Kaldor, J. & Clayton, D. (1985). Latent class analysis in chronic disease epidemiology. *Statistics in Medicine* **4**, 327–35.

Kauman, W.G., Gottstein, J.W. & Lantican, D. (1956). Quality evaluation by numerical and subjective methods with application to dried veneer. *Biometrics* **12**, 127–53.

Kelly, G.E. (1985). Use of the structural equations model in assessing the reliability of a new measurement technique. *Applied Statistics* **34**, 258–63.

Kendall, M.G. (1943). *The Advanced Theory of Statistics* (5th Edition). London: Griffin.

Kenny, D.A. & Kashby, D.A. (1992). Analysis of the multitrait-multimethod matrix by confirmatory factor analysis. *Psychological Bulletin* **112**, 165–72.

Kim, M.G. (2000). Outliers and influential observations in the structural errors-in-variables model. *Journal of Applied Statistics* **27**, 451–60.

Kimura, D.K. (1992). Functional comparative calibration using an EM algorithm. *Biometrics* **48**, 1263–71.

King, T.S. & Chinchilli, V.M. (2001). A generalized concordance correlation coefficient for continuous and categorical data. *Statistics in Medicine* **20**, 2131–47.

Kipnis, V., Carroll, R.J., Freedman, L.S. & Li, L. (1999). Implications of a new dietary measurement error model for estimation of relative risk: application to four calibration studies. *American Journal of Epidemiology* **150**, 642–51.

Kipnis, V., Midthune, D., Freedman, L.S., Bingham, S., Schatzkin, A., Subar, A. & Carroll, R.J. (2001). Empirical evidence of correlated biases in dietary assessments and its implications. *American Journal of Epidemiology* **153**, 394–403.

Klar, N., Lipsitz, S.R. & Ibrahim, J.G. (2000). An estimating equations approach for modelling kappa. *Biometrical Journal* **42**, 45–58.

Klar, N., Lipsitz, S.R., Parzen, M. & Leong, T. (2002). An exact bootstrap confidence interval for κ in small samples. *The Statistician* **51**, 467–78.

Koch, G.G, Landis, J.R., Freeman, J.L., Freeman, D.H. & Lehnen, R.G. (1977). A general methodology for the analysis of experiments with repeated measurement of categorical data. *Biometrics* **33**, 133–58.

Konishi, S. (1985). Normalizing and variance stabilizing transformations for intraclass correlations. *Annals of the Institute of Statistical Mathematics* **37**, 87–94.

Konishi, S. & Gupta, A.K. (1989). Testing the equality of several intraclass correlation coefficients. *Journal of Statistical Planning and Inference* **21**, 93–105.

Kosinski, A.J. & Barnhart, H.X. (2003). Accounting for nonignorable verification bias in assessment of diagnostic tests. *Biometrics* **59**, 163–71.

Koval, J.J. & Blackman, N.J.-M. (1996). Estimators of kappa – exact small sample properties. *Journal of Statistical Computation & Simulation* **55**, 315–36.

Kraemer, H.C. (1979). Ramifications of a population model for κ as a coeffcient of reliability. *Psychometrika* **44**, 461–71.

Kraemer, H.C. (1980). Extension of kappa coefficient. *Biometrics* **36**, 207–16.

Kraemer, H.C. (1981). Extension of Feldt's approach to testing homogeneity of coefficients of reliability. *Psychometrika* **46**, 41–5.

Kraemer, H.C. (1988). Assessment of 2×2 associations: generalization of signal-detection methodology. *The American Statistician* **42**, 37–49.

Kraemer, H.C. (1992a). *Statistical Evaluation of Medical Tests*. Thousand Oaks, CA: Sage.

Kraemer, H.C. (1992b). Measurement of reliability for categorical data in medical research. *Statistical Methods in Medical Research* **1**, 183–99.

Kraemer, H.C. & Bloch, D.A. (1988). Kappa coefficients in epidemiology: an appraisal of a reappraisal. *Journal of Clinical Epidemiology* **41**, 959–68.

Kraemer, H.C. & Bloch, D.A. (1990). A note on case-control sampling to estimate kappa coeffcients. *Biometrics* **46**, 49–59.

Kraemer, H.C., Periyakoil, V.S. & Noda, A. (2002). Kappa coefficients in medical research. *Statistics in Medicine* **21**, 2109–29.

Krippendorff, K. (1970). Bivariate agreement coefficients for reliability of data. In E.F. Borgatta & G.W. Bohrnstedt (Eds), *Sociological Methodology*, 139–50. San Francisco: Jossey Bass.

Krippendorff, K. (1980). *Content Analysis*. Beverly Hills, CA: Sage.

Kristiansen, S. (1991). A statistical procedure for the estimation of accuracy parameters in interlaboratory studies. *Statistics in Medicine* **10**, 843–54.

Kristof, W. (1969). Estimation of true score and error variance for tests under various equivalence assumptions. *Psychometrika* **34**, 489–507.

Krummenauer, F. (1999). Intraindividual scale comparison in clinical diagnostic methods: a review of elementary methods. *Biometrical Journal* **41**, 917–29.

Krummenauer, F. & Doll, G. (2000). Statistical methods for the comparison of measurements from orthodontic imaging. *European Journal of Orthodontics* **22**, 257–69.

Krummenauer, F., Genevrière, I. & Nixdorff, U. (2000). The biometrical comparison of cardiac imaging methods. *Computer Methods and Programs in Biomedicine* **62**, 21–34.

Kummel, C.H. (1879). Reduction of observation equations which contain more than one observed quantity. *The Analyst* **6**: 97–105.

Kuttatharmmakul, S., Massart, D.L. & Smeyers-Verbeke, J. (2000). Comparison of alternative measurement methods: determination of the minimal number of measurements required for the evaluation of bias by means of interval hypothesis testing. *Chemometrics and Intelligent Laboratory Systems* **52**, 61–73.

Lakshminarayanan, M.Y. and Gunst, R.F. (1984). Estimation of the parameters in linear structural relationships: sensitivity to choice of the ratio of error variances. *Biometrika* **71**, 569–73.

LaMotte, L.R. (1973). Quadratic estimation of variance components. *Biometrics* **29**, 311–30.

Landis, J.R. & Koch, G.G. (1975). A review of statistical methods on the analysis of data arising from observer reliability studies. *Statistica Neerlandica* **29**, 101–23 and 151–61.

Landis, J.R. & Koch, G.G. (1977a). The measurement of observer agreement for categorical data. *Biometrics* **33**, 159–74.

Landis, J.R. & Koch, G.G. (1977b). An application of hierarchical kappa-type statistics in the assessment of majority agreement among multiple raters. *Biometrics* **33**, 363–74.

Landis, J.R. & Koch, G.G. (1977c). A one-way components of variance model for categorical data. *Biometrics* **33**, 671–9.

Langeheine, R., Pannekoek, J. & Van de Pol, F. (1996). Bootstrapping goodness-of-fit measures in categorical data analysis. *Sociological Methods & Research* **24**, 492–516.

Lau, T.-S. (1997). The latent class model for multiple binary screening tests. *Statistics in Medicine* **16**, 2283–95.

Lavori, P.W. & Keller, M.B. (1988). Improving the aggregate performance of psychiatric diagnostic methods when not all subjects receive the standard test. *Statistics in Medicine* **7**, 727–37.

Lawton, W.H., Sylvestre, E.A. & Young-Ferraro, B.J. (1979). Statistical comparison of multiple analytic procedures: application to clinical chemistry. *Technometrics* **21**, 397–409.

Lazarsfeld, P.F. & Henry, N.W. (1968). *Latent Structure Analysis*. New York: Houghton-Mifflin.

Lee, T.D. (1987). Assessment of inter- and intra-laboratory variances: a Bayesian alternative to BS 5497. *The Statistician* **36**, 161–70.

Leisenring, W., Alonzo, T. & Sullivan Pepe, M. (2000). Comparisons of predictive values of binary medical diagnostic tests for paired designs. *Biometrics* **56**, 345–51.

Levin, J. (1993). Stability coefficients in longitudinal studies. *Applied Psychological Measurement* **17**, 17–19.

Lewin-Koh, S.-C. & Amemiya, Y. (2003). Heteroscedastic factor analysis. *Biometrika* **90**, 85–97.

Lewis, G., Pelosi, A.J., Araya, R. & Dunn, G. (1992). Measuring psychiatric disorder in the community: a standardised assessment for use by lay interviewers. *Psychological Medicine* **22**, 465–86.

Lewis, P.A., Jones, P.W., Polak, J.W. & Tillitson, H.T. (1991). The problem of conversion in method comparison studies. *Applied Statistics* **40**, 105–12.

Liang, K.-Y. & Zeger, S.L. (1986). Longitudinal data analysis using generalized linear models. *Biometrika* **73**, 13–22.

Light, R.J. (1971). Measures of response agreement for qualitative data: some generalizations and alternatives. *Psychological Bulletin* **76**, 365–77.

Lijmer, J.G., Bossuyt, P.M.M. & Heisterkamp, S.H. (2002). Exploring sources of heterogeneity in systematic reviews of diagnostic tests. *Statistics in Medicine* **21**, 1525–37.

Lin, L., Hedayat, A.S., Sinha, B. & Yang, M. (2002). Statistical methods in assessing agreement: models, issues and tools. *Journal of the American Statistical Association* **97**, 257–70.

Lin, L.I.-K. (1989). A concordance correlation coefficient to evaluate reproducibility. *Biometrics* **45**, 255–68 (see Corrections in *Biometrics* **56**, 324–5).

Lin, L.I.-K. (1992). Assay validation using the concordance correlation coefficient. *Biometrics* **48**, 599–604.

Lin, L.I.-K. (2000). Total deviation index for measuring individual agreement with applications in laboratory performance and bioequivalence. *Statistics in Medicine* **19**, 255–70.

Linnet, K. (1990). Estimation of linear relationship between measurements of two methods with proportional errors. *Statistics in Medicine* **9**, 1463–73.

Linnet, K. (1998). Performance of Deming regression analysis in case of misspecified analytical error ratio in method comparison studies. *Clinical Chemistry* **44**, 1024–31.

Linnet, K. (1999). Necessary sample size for method comparison studies based on regression analysis. *Clinical Chemistry* **45**, 882–94.

Lipsitz, S.R., Laird, N.M. & Brennan, T.A. (1994). Simple moment estimates of the κ-coefficient and its variance. *Applied Statistics* **43**, 309–23.

Lipsitz, S.R., Laird, N.M., Brennan, T.A. & Parzen, M. (2001). Estimating the κ-coefficient from a selected sample. *The Statistician* **50**, 407–16.

Little, R.J.A. & Rubin, D.B. (2002). *Statistical Analysis with Missing Data* (2nd Edition). New York: Wiley.

Llabre, M.M., Ironson, G.H., Spitzer, S.B., Gellman, M.D., Weidler, D.J. & Schneiderman, N. (1988). How many blood pressure measurements are enough?: an application of generalizability theory to the study of blood pressure reliability. *Psychophysiology* **25**, 97–106.

Lu, Y., Ye, K., Mathur, A.K., Hui, S., Fuerst, T.P. & Genant, H. (1997). Comparative calibration without a gold standard. *Statistics in Medicine* **16**, 1889–905.

Ludbrook, J. (1997). Comparing methods of measurement. *Clinical and Experimental Pharmacology and Physiology* **24**, 193–203.

Lui, K.-J. & Kelly, C. (1999). A note on interval estimation of kappa in a series of 2×2 tables. *Statistics in Medicine* **18**, 2041–9.

Lyles, R.H. & Chambless, L.E. (1995). Effects of model misspecification in the estimation of variance components and intraclass correlation for paired data. *Statistics in Medicine* **14**, 1693–706.

Lyles, R.H., Fan, D. & Chuachoowong, R. (2001). Correlation coefficient estimation involving a left censored laboratory assay variable. *Statistics in Medicine* **20**, 2921–33.

MacFarlane, R.G., King, E.J., Wootton, I.D.P. & Gilchrist, M. (1948). Determination of haemoglobin III. Reliability of clinical and other methods. *Lancet* **i**, 282–6.

Macnab, A.J., Levine, M., Glick, N., Phillips, N., Susak, L. & Elliott, M. (1994). The Vancouver sedative recovery scale for children: validation and reliability of scoring based on videotaped instruction. *Canadian Journal of Anaesthesia* **41**, 913–18.

Magidson, J. & Vermunt, J.K. (2000). *Latent GOLD*. Belmont, MA: Statistical Innovations.

Magidson, J. & Vermunt, J.K. (2003). Latent class analysis. In D. Kaplan (Ed.), *Handbook of Quantitative Methodology for the Social Sciences*, in press. Beverly Hills: Sage.

Mak, T.K. (1988). Analysing intraclass correlation for dichotomous variables. *Applied Statistics* **37**, 344–52.

Maloney, C.J. & Rastogi, S.C. (1970). Significance test for Grubbs's estimators. *Biometrics* **26**, 671–6.

Mandel, J. (1959). The measuring process. *Technometrics* **1**, 251–67.

Mandel J. (1964). *The Statistical Evaluation of Experimental Data*. New York: Wiley & Sons.

Mandel, J. (1991). *Evaluation and Control of Measurements*. New York: Marcel Dekker Inc.

Mandel, J. (1995). *The Analysis of Two-Way Layouts*. New York: Wiley & Sons.

Marcoulides, G.A. (1993). Maximizing power in generalizability studies under budget constraints. *Journal of Educational Statistics* **18**, 197–206.

Marcoulides, G.A. (1995). Designing measurement studies under budget constraints: controlling error of measurement and power. *Educational and Psychological Measurement* **55**, 423–8.

Markus, G.B. (1979). *Analysing Panel Data*. Beverly Hills & London: Sage Publications.

Martin, R.F. (2000). General Deming regression for estimating systematic bias and its confidence interval in method-comparison studies. *Clinical Chemistry* **46**, 100–4.

Martus, P. (2000). Statistical methods for the evaluation of diagnostic measurements concerning paired organs. *Statistics in Medicine* **19**, 525–40.

Martus, P. (2001). A measurement model for disease in paired organs. *Biometrical Journal* **43**, 927–40.

Matthews, J.N.S. (1997). A formula for the probability of discordant classification in method comparison studies. *Statistics in Medicine* **16**, 705–10.

Maxwell, A.E. & Pilliner, A.E.G. (1968). Deriving coefficients of reliability and agreement for ratings. *British Journal of Mathematical and Statistical Psychology* **21**, 105–16.

McLachlan, G.J. & Peel, D. (2000). *Finite Mixture Models*. New York: Wiley & Sons.

McNemar, Q. (1947). Note on the sampling error of the difference between correlated proportions or percentages. *Psychometrika* **12**, 153–7.

Meijer, E. & Mooijaat, A. (1996). Factor analysis with heteroscedastic errors. *British Journal of Mathematical and Statistical Psychology* **49**, 189–202.

Messer, S.C. & Gross, A.M. (1994). Childhood depression and aggression: a covariance structure analysis. *Behavior Research and Therapy* **32**, 633–77.

Meyers, H. & von Hake, C.A. (1976). *Earthquake Data File Summary*. National Geophysical and Solar-Terrestrial Data Center, US Department of Commerce, Boulder, Colorado.

Mian, I.U.H. & Shoukri, M.M. (1997). Statistical analysis of intraclass correlations from multiple samples with applications to arterial blood pressure data. *Statistics in Medicine* **16**, 1497–514.

Mooijaart, A. (1985). Factor analysis for non-normal variables. *Psychometrika* **50**, 323–42.

Morgan, W.A. (1939). A test for the significance of the difference between the two variances in a sample from a normal bivariate population. *Biometrika* **31**, 13–19.

Morton, A.P. & Dobson, A.J. (1989). Assessing agreement. *Medical Journal of Australia* **150**: 384–7.

Mosteller, F. & Tukey, J.W. (1977). *Data Analysis and Regression*. New York: Addison-Wesley.

Müller, R. & Büttner, P. (1995). A critical discussion of intraclass correlation coefficients. *Statistics in Medicine* **13**, 2465–76.

Muthén, B. (1989). Factor structure in groups selected on observed scores. *British Journal of Mathematical and Statistical Psychology* **42**, 81–90.

Muthén, B. & Kaplan, D. (1992). A comparison of some methodologies for the factor analysis of non-normal Likert variables: a note on the size of the model. *British Journal of Mathematical and Statistical Psychology* **45**, 19–30.

Muthén, B., Kaplan, D. & Hollis, M. (1987). On structural equation modelling with data that are not missing completely at random. *Psychometrika* **52**, 431–62.

Muthén, L.K. & Muthén, B.O. (1998–2001). *Mplus User's Guide*. Los Angeles, CA: Muthén & Muthén.

Muthén, B. & Sheddon, K. (1999). Finite mixture modelling with mixture outcomes using the EM algorithm. *Biometrics* **55**, 463–9.

Nagelkerke, N., Fidler, V. & Buwalda, M. (1988). Instrumental variables in the evaluations of diagnostic test procedures when the true disease status is unknown. *Statistics in Medicine* **7**, 739–44.

Nam, J.-M. (2000). Interval estimation of the kappa coefficient with binary classification and an equal marginal probability model. *Biometrics* **56**, 583–5.

Nam, J.-M. (2002). Testing the intraclass version of kappa coefficient of agreement with binary scale and sample size determination. *Biometrical Journal* **44**, 558–70.

Napoli, K.L., Chan, W. & Wang, M.-E. (1996). An appropriate statistical approach for the comparison of FPIA to HPLC for determination of cyclosporine concentrations in animal tissues. *Journal of Clinical Ligand Assay* **19**, 190–7.

Nelson, J.C. & Sullivan Pepe, M. (2000). Statistical description of interrater variability in ordinal ratings. *Statistical Methods in Medical Research* **9**, 475–96.

Nickerson, C.A.E. (1997). A note on 'A concordance correlation coefficient to evaluate reproducibility'. *Biometrics* **53**, 1503–7.

Nix, A.B.J. & Dunstan, F.D.J. (1991). Maximum likelihood techniques applied to method comparison studies. *Statistics in Medicine* **10**: 981–8.

Nurminen, M., Hytönen, M. & Sala, E. (2000). Modelling the reproducibility of acoustic rhinometry. *Statistics in Medicine* **19**, 1179–89.

Obuchowski, N.A. & Zhou, X.-H. (2002). Prospective studies of diagnostic test accuracy when disease prevalence is low. *Biostatistics* **3**, 477–92.

Oden, N.L. (1991). Estimating kappa from binocular data. *Statistics in Medicine* **10**, 1303–11.

O'Grady, K.E. & Medoff, D.R. (1991). Rater reliability: a maximum likelihood confirmatory factor-analytic approach. *Multivariate Behavioral Research* **26**, 363–87.

Olin, B.D. & Meeker, W.Q. (1996). Applications of statistical methods to non-destructive evaluation. *Technometrics* **38**, 95–112.

Oman, S.D., Meir, N. & Haim, N. (1999). Comparing two measures of creatinine clearance: an application of errors-in-variables and bootstrap techniques. *Applied Statistics* **48**, 39–52.

Patefield, M. (2002). Fitting non-linear structural relationships using SAS procedure NLMIXED. *The Statistician* **51**, 355–66.

Patterson, H.D. & Thompson, R. (1971). The recovery of inter-block information when block sizes are unequal. *Biometrika* **58**, 545–54.

Patterson, H.D. & Thompson, R. (1975). Maximum likelihood estimation of components of variance. *Proceedings of the 8th International Biometric Conference*, 197–207.

Paul, S.R. & Barnwal, R.K. Maximum likelihood estimation and a C(α) test for a common intraclass correlation. *The Statistician* **39**, 19–24.

Pearson, K. (1901). On lines and planes of closest fit to systems of points in space. *Philosophical Magazine* **2**, *Series 6*, 559–72.

Perisic, I. & Rosner, B. (1999). Comparisons of measures of interclass correlations: the general case of unequal group size. *Statistics in Medicine* **18**, 1451–66.

Permutt, T., Edland, S.D., Moezzi, M. & Grosser, S.C. (1991). Likelihood techniques for interlaboratory calibration in the National Stream Survey. *Journal of Chemometrics* **5**, 299–308.

Pitman, E.J.G. (1939). A note on normal correlation. *Biometrika* **31**, 9–12.

Plummer, M. & Clayton, D. (1993a). Measurement error in dietary assessment: an investigation using covariance structure models: Part I. *Statistics in Medicine* **12**, 925–35.

Plummer, M. & Clayton, D. (1993b). Measurement error in dietary assessment: an investigation using covariance structure models: Part II. *Statistics in Medicine* **12**, 937–48.

Plummer, M., Clayton, D. & Kaaks, R. (1994). Calibration in multi-centre cohort studies. *International Journal of Epidemiology* **23**, 419–26.

Qu, Y.S. & Hadgu, A. (1998). A model for evaluating sensitivity and specificity for correlated diagnostic tests in efficacy studies with an imperfect reference test. *Journal of the American Statistical Association* **93**, 920–8.

Qu, Y.S., Tan, M. & Kutner, M.H. (1996). Random effects models in latent class analysis for evaluating accuracy of diagnostic tests. *Biometrics* **52**, 797–810.

Quan, H. & Shih, W.J. (1996). Assessing reproducibility by a within-subject coefficient of variation with random effects models. *Biometrics* **52**, 1195–203.

Rabe-Hesketh, S., Skrondal, A. & Pickles, A. (2002). Reliable estimation of generalized linear mixed models using adaptive quadrature. *Stata Journal* **2**, 1–21.

Rabe-Hesketh, S., Skrondal, A. & Pickles, A. (2004). Generalized multilevel structural equation modelling. *Psychometrika*, in press.

Rae, G. (1984). On measuring agreement among several judges on the presence or absence of a trait. *Educational and Psychological Measurement* **44**, 247–53.

Rae, G. (1988). The equivalence of multiple rater kappa statistics and intraclass correlation coefficients. *Educational and Psychological Measurement* **48**, 367–74.

Raffalovitch, L.E. & Bohrnstedt, G.W. (1987). Common, specific and error variance components of factor models: estimation with longitudinal data. *Sociological Methods & Research* **15**, 385–405.

Rao, C.R. (1972). Estimation of variance and covariance components in linear models. *Journal of the American Statistical Association* **67**, 112–15.

Read, T.R.C. & Cressie, N.A.C. (1988). *Goodness-of-Fit Statistics for Discrete Multivariate Data*. New York: Springer-Verlag.

Reiser, M. & Lin, Y. (1999). A goodness-of-fit test for the latent class model when expected frequencies are small. In Sobel, M.E. & Becker, M.P. (Eds), *Sociological Methodology 1999*, 81–111.

Richardson, S., Leblond, L., Jaussent, I. & Green, P.J. (2002). Mixture models in measurement error problems, with reference to epidemiological studies. *Journal of the Royal Statistical Society, Series A* **165**, 549–66.

Ridout, M.S., Demétrio, C.G.B. & Firth, D. (1999). Estimating intraclass correlation for binary data. *Biometrics* **55**, 137–48 (Correction **59**, 199 (2003)).

Rindskopf, D. & Rindskopf, W. (1986). The value of latent class analysis in medical diagnosis. *Statistics in Medicine* **5**, 21–7.

Roberts, C. & McNamee, R. (1998). A matrix of kappa-type coefficients to assess the reliability of nominal scales. *Statistics in Medicine* **17**, 471–88.

Robinson, D.L. (1987). Estimation and use of variance components. *The Statistician* **36**, 3–14.

Rocke, D.M. (1983). Robust statistical analysis of interlaboratory studies. *Biometrika* **70**, 421–31.

Rocke, D. & Lorenzato, S. (1995). A two-component model for measurement error in analytical chemistry. *Technometrics* **37**, 176–84.

Rodbard, D. (1978). Statistical estimation of the minimum detectable concentration ('sensitivity') for radioligand assays. *Analytical Biochemistry* **90**, 1–12.

Rofel, A., Boëlle, P.Y. & Mary, J.Y. (1998). Global and partial agreement among several observers. *Statistics in Medicine* **17**, 489–501.

Rubin, D.B. & Thayer, D.T. (1982). EM algorithms for ML factor analysis. *Psychometrika* **47**, 69–76.

Rubin, D.B. & Thayer, D.T. (1983). More on EM for ML factor analysis. *Psychometrika* **48**, 253–7.

Russell, T.S. & Bradley, R.A. (1958). One-way variances in a two-way classification. *Biometrika* **45**, 111–29.

Särndal, C.-E., Swensson, B. & Wretman, J.(1992). *Model Assisted Survey Sampling*. Berlin & New York: Springer.

Satterthwaite, F.E. (1946). An approximate distribution of estimates of variance components. *Biometrics Bulletin* **2**, 110–4.

Satorra, A. & Bentler, P.M. (1990). Model conditions for asymptotic robustness in the analysis of linear relations. *Computational Statistics and Data Analysis* **10**, 235–49.

Scott, W.A. (1955). Reliability of content analysis: the case of nominal scale coding. *Public Opinion Quarterly* **19**, 321–5.

Schemper, M. (1984). A generalization of the intraclass tau correlation for tied and censored data. *Biometrical Journal* **26**, 609–17.

Schemper, M. (1986). General derivation of intraclass correlation coefficients. *Biometrical Journal* **28**, 485–9.

Schouten, H.J.A. (1985). *Statistical Measurement of Interobserver Agreement*. Unpublished doctoral dissertation. Erasmus University, Rotterdam.

Schouten, H.J.A. (1986). Nominal scale agreement among observers. *Psychometrika* **51**, 453–66.

Schouten, H.J.A. (1993). Estimating kappa from binocular data and comparing marginal probabilities. *Statistics in Medicine* **12**, 2207–17.

Schuster, C. & von Eye, A. (2001). Models for ordinal agreement data. *Biometrical Journal* **43**, 795–808.

Scott, W.A. (1955). Reliability of content analysis: the case for nominal scale coding. *Public Opinion Quarterly* **19**, 321–5.

Searle, S.R. (1987). *Linear Models for Unbalanced Data*. New York: Wiley.

Searle, S.R., Casella, G. & McCulloch, C.E. (1992). *Variance Components*. New York: Wiley.

Shao, J. & Tu, D. (1995) *The Jackknife and the Bootstrap*. New York: Springer-Verlag.

Shavelson, R.J. & Webb, N.M. (1991). *Generalizability Theory: A Primer*. Newbury Park, CA: Sage.

Shen, Y., Wu, D. & Zelen, M. (2001). Testing the independence of two diagnostic tests. *Biometrics* **57**, 1009–17.

Shirahata, S. (1982). Nonparametric measures of intraclass correlation. *Communications in Statistics – Theory and Methods* **11**, 1707–21.

Shirahata, S. (1982). A nonparametric measure of intraclass correlation. *Communications in Statistics – Theory and Methods* **11**, 1723–32.

Shoukri, M.M. (2003). *Measurement of Agreement for Categorical Data.* London: Chapman & Hall/CRC.

Shoukri, M.M. & Demirkaya, O. (2000). Sample size requirements to test the equality of rater's precision. *Journal of Applied Statistics* **27**, 483–94.

Shoukri, M.M. & Donner, A. (2001). Efficiency considerations in the analysis of inter-observer agreement. *Biostatistics* **2**, 323–36.

Shrout, P.E. (1998) Measurement reliability and agreement in psychiatry. *Statistical Methods in Medical Research* **7**, 301–17.

Shrout, P.E. & Fleiss, J.L. (1979). Intra-class correlations: uses in assessing rater reliability. *Psychological Bulletin* **86**, 420–8.

Shukla, G.K. (1972). An invariant test for the homogeneity of variances in a two-way classification. *Biometrics* **28**, 1063–72.

Shukla, G.K. (1973). Some exact tests of hypotheses about Grubbs' estimators. *Biometrics* **29**, 373–7.

Shyr, J.Y. & Gleser, L.J. (1986). Inference about comparative precision in linear structural relationships. *Journal of Statistical Planning and Inference* **14**, 339–58.

Simonoff, E., Pickles, A., Hewitt, J., Silberg, J., Rutter, M., Loeber, R., Meyer, J., Neale, M. & Eaves, L. (1995). Multiple raters of disruptive child behavior: using a genetic strategy to examine shared views and bias. *Behavior Genetics* **25**, 311–26.

Sinclair, M.D. & Gastwith, J.L. (1996). On procedures for evaluating the effectiveness of reinterview survey methods: application to labor force data. *Journal of the American Statistical Association* **91**, 961–9.

Singh, B. (2003). Estimation of heritability in heteroscedastic one-way unbalanced random model. *Biometrical Journal* **45**, 527–40.

Snedecor, G.W. & Cochran, W.G. (1967). *Statistical Methods* (6th Edition). Ames: Iowa State University.

Sörbom, D. (1982). Structural equation models with structured means. In K.G. Jöreskog & H. Wold (Eds), *Systems under Indirect Observation: Causality, Structure, Prediction* 183–95. Amsterdam: North Holland.

Spearman, C. (1904). General intelligence objectively determined and measured. *American Journal of Psychology* **15**, 201–93.

St Laurant, R.T. (1998). Evaluating agreement with a gold standard in method comparison studies. *Biometrics* **54**, 5237–545.

StataCorp (2003). *Stata Statistical Software: Release 8.0.* College Station, TX: Stata Corporation.

Stavig, G.R. (1984). Monotonic measures of agreement for ranked data. *British Journal of Mathematical and Statistical Psychology* **37**, 283–7.

Steichen, D.J. & Cox, N.J. (2002). A note on the concordance correlation coefficient. *The Stata Journal* **2**, 183–9.

Strike, P.W. (1991). *Statistical Methods in Laboratory Medicine.* Oxford: Butterworth-Heinemann.

Strike, P.W. (1995). *Measurement in Laboratory Medicine.* Oxford: Butterworth-Heinemann.

Sullivan Pepe, M. (2003). *The Statistical Evaluation of Medical Tests for Classification and Prediction.* Oxford: Oxford University Press.

Sullivan Pepe, M. & Alonzo, T.A. (2001). Comparing disease screening tests when true disease status is ascertained only for screen positives. *Biostatistics* **2**, 249–60.

Svensson, E. (2000). Comparison of the quality of assessments using continuous and discrete ordinal rating scales. *Biometrical Journal* **42**, 417–34.

Svensson, E. (2000). Concordance between ratings using different scales for the same variable. *Statistics in Medicine* **19**, 3483–96.

Tan, C.Y. & Iglewicz, B. (1999). Measurement-methods comparisons and linear structural relationship. *Technometrics* **41**, 192–201.

Tanner, M.A. & Young, M.A. (1985). Modeling ordinal scale disagreement. *Psychological Bulletin* **98**, 408–15.

Tanner, M.A. & Young, M.A. (1985). Modeling agreement among raters. *Journal of the American Statistical Association* **80**, 175–80.

Taylor, J.R. (1982). *An Introduction to Error Analysis*. Mill Valley, CA: Oxford University Press.

Theobald, C.M. & Mallinson, J.R. (1978). Comparative calibration: linear structural relationships and congeneric measurements. *Biometrics* **34**, 39–45.

Thompson, J.R. (2001). Estimating equations for kappa statistics. *Statistics in Medicine* **20**, 2895–906.

Thompson, W.A. Jr. (1963). Precision of simultaneous test procedures. *Journal of the American Statistical Association* **58**, 474–9.

Thürigen, D., Spiegelman, D., Blettner, M., Heuer, C. & Brenner, H. (2000). Measurement error correction using validation data: a review of methods and their applicability in case-control studies. *Statistical Methods in Medical Research* **9**, 447–74.

Titterington, D.M., Smith, A.F.M. & Markov, U.E. (1985). *Statistical Analysis of Finite Mixture Distributions*. Wiley: New York.

Torrance-Rynard, V.L. & Walter, S.D. (1997). Effects of dependent errors in the assessment of diagnostic test performance. *Statistics in Medicine* **16**, 2157–75.

Tosteson, T.D., Titus-Ernstoff, L., Baron, J.A. & Karaglas, M.R. (1994). A two-stage validation study for determining sensitivity and specificity. *Environmental Health Perspectives* **102** (Suppl. 8), 11–15.

Traub, R.E. (1994). *Reliability for the Social Sciences*. Thousand Oaks, CA: Sage.

Turner, S.W., Toone, B.K. & Brett-Jones, J.R. (1986). Computerized tomographic scan changes in early schizophrenia – preliminary findings. *Psychological Medicine* **16**, 219–25.

Uebersax, J.S. & Grove, W.M. (1990). Latent class analyis of diagnostic agreement. *Statistics in Medicine* **9**, 559–72.

Uebersax, J.S. & Grove, W.M. (1993). A latent trait finite mixture model for the analysis of rating agreement. *Biometrics* **49**, 823–35.

Vacek, P.M. (1985). The effect of conditional dependence on the evaluation of diagnostic tests. *Biometrics* **41**, 959–68.

Valenstein, P.N. (1990). Evaluating diagnostic tests with imperfect standards. *American Journal of Clinical Pathology* **93**, 252–8.

Vangel, M.G. & Rukhin, A.L. (1999). Maximum likelihood analysis for heteroscedastic one-way random effects ANOVA in interlaboratory studies. *Biometrics* **55**, 129–36.

Vanleeuwen, D.M. & Mandabach, K.H. (2002). A note on the reliability of ranked items. *Sociological Methods & Research* **31**, 87–105.

Vargha, P. (1997). Letter to the Editor. Re: 'A critical discussion of intraclass correlation coefficients' by R. Müller & P. Büttner in *Statistics in Medicine* **13**, 2465–76 (1995). *Statistics in Medicine* **16**, 821–3.

Voss, B., Kunert, J., Dahms, S. & Weiss, H. (2000). A multinomial model for the quality control of colony counting procedures. *Biometrical Journal* **42**, 263–78.

Wacholder, S., Armstrong, B. & Hartge, P. (1993). Validation studies using an alloyed gold standard. *American Journal of Epidemiology* **137**, 1251–8.

Walter, S.D. (1984). Measuring the reliability of clinical data: the case for using three observers. *Revue Epidémiologie et Santé Publique* **32**, 206–11.

Walter, S.D. (1999). Estimation of test sensitivity and specificity when disease confirmation is limited to positive results. *Epidemiology* **10**, 67–72.

Walter, S.D., Eliasziw, M. & Donner, A. (1998). Sample size and optimal study designs for reliability studies. *Statistics in Medicine* **17**, 101–10.

Walter, S.D. & Irwig, L.M. (1988). Estimation of test error rates, disease prevalence and relative risk from misclassified data: a review. *Journal of Clinical Epidemiology* **41**, 923–37.

Walter, S.D., Irwig, L.M. & Glasziou, P.P. (1999). Meta-analysis of diagnostic tests with imperfect reference standards. *Journal of Clinical Epidemiology* **52**, 943–51.

Wang, C.Y. (2000). Weighted normality-based estimator in correcting correlation coefficient estimation between nutrient measurements. *Biometrics* **56**, 106–12.

Warton, D.I. & Weber, N.C. (2002). Common slope tests for bivariate errors-in-variables models. *Biometrical Journal* **44**, 161–74.

Wechsler, D. (1981). *WAIS-R Manual*. New York: Psychological Corporation.

Wedel, M. (2002). Concomitant variables in finite mixture models. *Statistica Neerlandica* **56**, 362–75.

Wehrens, R., Putter, H. & Buydens, L.M.C. (2000). The bootstrap: a tutorial. *Chemometrics and Intelligent Laboratory Systems* **54**, 35–52.

Werts, C.E., Rock, D.R. & Grandy, J. (1979). Confirmatory factor analysis applications: missing data problems and comparison of path models between populations. *Multivariate Behavioral Research* **14**, 199–213.

Wheaton, B., Muthén, B., Alwin, D.F. & Summers, G.F. (1977). Assessing reliability and stability in panel models. In D.R. Heise (Ed.), *Sociological Methodology 1977*, 84–136. San Francisco: Jossey Bass.

White, H. (1982). Maximum likelihood estimation of mis-specified models. *Econometrics* **50**, 1–25.

Wiley, E.W. (2000). *Bootstrap strategies for variance components estimation: theoretical and empirical results*. Unpublished doctoral dissertation, Stanford (cited in Brennan, 2001).

Williams, E.J. (1952). The interpretation of interactions in factorial experiments. *Biometrika* **39**, 65–81.

Williams, E.J. (1959). *Regression Analysis*. New York: Wiley & Sons.

Williams, E.J. (1973). Tests of correlation in multivariate analysis. *Proceedings of the 39th Session of the International Statistical Institute* **45**, book 4, 218–32.

Williams, G.W. (1976). Comparing the joint agreement of several raters with another rater. *Biometrics* **32**, 619–27.

Williamson, J.M. & Manatunga, A.K. (1997). Assessing interrater agreement from dependent data. *Biometrics* **53**, 707–14.

Williamson, J.M., Manatunga, A.K. & Lipsitz, S.R. (2000). Modeling kappa for measuring dependent categorical agreement data. *Biostatistics* **1**, 191–202.

Williamson, P.R., Lancaster, G.A., Craig, J.V. & Smyth, R.L. (2002). Mata-analysis of method comparison studies. *Statistics in Medicine* **21**, 2013–25.

Wong, M.Y., Day, N.E., Bashir, S.A. & Duffy, S.W. (1999). Measurement error in epidemiology: the design of validation studies I: univariate situation. *Statistics in Medicine* **18**, 2815–29.

Wong, M.Y., Day, N.E. & Wareham, N.J. (1999). Measurement error in epidemiology: the design of validation studies II: multivariate situation. *Statistics in Medicine* **18**, 2831–45.

Wooldridge, J.M. (2000). *Introductory Econometrics: A Modern Approach*. United States: South-Western College Publishing.

Wothke, W. (1996). Models for multitrait-multimethod matrix analysis. In G.A. Marcoulides & R.E. Schumaker (Eds), *Advanced Structural Equation Modeling: Issues and Techniques*, 7–56. Mahwah, N.J.: Lawrence Erlbaum.

Yamamoto, E. & Yanagimoto, T. (1992). Moment estimators for the binomial distribution. *Journal of Applied Statistics* **19**, 273–83.

Yang, I. & Becker, M. (1997). Latent variable modelling of diagnostic accuracy. *Biometrics* **53**, 948–58.

Yuan, K.-H. & Bentler, P.M. (1998). Structural equation modelling with robust covariances. In A.E. Raftery (Ed.), *Sociological Methodology* **28**, 363–96. Boston & Oxford: Blackwell.

Yuan, K.-H. & Bentler, P.M. (2000). Three likelihood-based methods for mean and covariance structure analysis with nonnormal missing data. In R.M. Stolzenberg (Ed.), *Sociological Methodology* **30**, 165–200. Boston & Oxford: Blackwell.

Yung, Y.-F. (1997). Finite mixtures in confirmatory factor-analysis models. *Psychometrika* **62**, 297–330.

Zeger, S.L. & Liang, K.-Y. (1986). Longitudinal data analysis for discrete and continuous outcomes. *Biometrics* **42**, 121–30.

Zeller, R.A. (1987). Comment on 'Common, specific and error variance components of factor models: estimation with longitudinal data'. *Sociological Methods & Research* **15**, 406–19.

Zhou, X.-H. (1998a). Comparing accuracies of two screening tests in a two-phase study of dementia. *Applied Statistics* **47**, 135–47.

Zhou, X.-H. (1998b). Correcting for verification bias in studies of a diagnostic test's accuracy. *Statistical Methods in Medical Research* **7**, 337–53.

Zhou, X.-H. & Higgs, R.E. (2000). Assessing the relative accuracies of two screening tests in the presence of verification biases. *Statistics in Medicine* **19**, 1697–705.

Zhou, X.-H., Obuchowski, N.A. & McClish, D.K. (2002). *Statistical Methods in Diagnostic Medicine*. New York: Wiley.

Zigmond, A.S. & Snaith, R.P. (1983). The Hospital Anxiety and Depression Scale. *Acta Psychiatrica Scandinavica* **67**, 361–70.

Zorn, M.E., Gibbons, R.D. & Sonzogni, W.C. (1997). Weighted least squares approach to calculating limits of detection and quantification by modelling variability as a function of concentration. *Analytical Chemistry* **69**, 3069–75.

Zou, K.H. & McDermott, M.P. (1999). Higher-moment approaches to approximate interval estimation for a certain intraclass correlation coefficient. *Statistics in Medicine* **18**, 2051–61.

Index

Printed and bound by CPI Group (UK) Ltd, Croydon, CR0 4YY

27/10/2024

14580284-0002